Mohammad Amin Rashidifar

Nonlinear Vibrations of Cantilever Beams and Plates

Anchor Academic
Publishing

Rashidifar, Mohammad Amin: Nonlinear Vibrations of Cantilever Beams and Plates,
Hamburg, Anchor Academic Publishing 2015

Buch-ISBN: 978-3-95489-420-8
PDF-eBook-ISBN: 978-3-95489-920-3
Druck/Herstellung: Anchor Academic Publishing, Hamburg, 2015

Bibliografische Information der Deutschen Nationalbibliothek:
Die Deutsche Nationalbibliothek verzeichnet diese Publikation in der Deutschen
Nationalbibliografie; detaillierte bibliografische Daten sind im Internet über
http://dnb.d-nb.de abrufbar.

Bibliographical Information of the German National Library:
The German National Library lists this publication in the German National Bibliography.
Detailed bibliographic data can be found at: http://dnb.d-nb.de

All rights reserved. This publication may not be reproduced, stored in a retrieval system
or transmitted, in any form or by any means, electronic, mechanical, photocopying,
recording or otherwise, without the prior permission of the publishers.

Das Werk einschließlich aller seiner Teile ist urheberrechtlich geschützt. Jede Verwertung
außerhalb der Grenzen des Urheberrechtsgesetzes ist ohne Zustimmung des Verlages
unzulässig und strafbar. Dies gilt insbesondere für Vervielfältigungen, Übersetzungen,
Mikroverfilmungen und die Einspeicherung und Bearbeitung in elektronischen Systemen.

Die Wiedergabe von Gebrauchsnamen, Handelsnamen, Warenbezeichnungen usw. in
diesem Werk berechtigt auch ohne besondere Kennzeichnung nicht zu der Annahme,
dass solche Namen im Sinne der Warenzeichen- und Markenschutz-Gesetzgebung als frei
zu betrachten wären und daher von jedermann benutzt werden dürften.

Die Informationen in diesem Werk wurden mit Sorgfalt erarbeitet. Dennoch können
Fehler nicht vollständig ausgeschlossen werden und die Diplomica Verlag GmbH, die
Autoren oder Übersetzer übernehmen keine juristische Verantwortung oder irgendeine
Haftung für evtl. verbliebene fehlerhafte Angaben und deren Folgen.

Alle Rechte vorbehalten

© Anchor Academic Publishing, Imprint der Diplomica Verlag GmbH
Hermannstal 119k, 22119 Hamburg
http://www.diplomica-verlag.de, Hamburg 2015
Printed in Germany

Table Of Contents

1	**Introduction**	**1**
	1.1 Motivation	1
	1.2 Types of Nonlinearity	2
	1.3 Literature Review	3
	1.3.1 Beam Theories	4
	1.3.2 Secondary Effects	7
	1.3.3 Modal Interactions	10
	1.3.4 System Identification	17
	1.3.5 Solution Methodologies	18
	1.4 Overview	20
2	**Problem Formulation**	**22**
	2.1 Beam Kinematics	22
	2.1.1 Euler-Angle Rotations	24
	2.1.2 Inextensional Beam	26

		2.1.3	Strain-Curvature Relations	27
	2.2	Equations of Motion		29
		2.2.1	Lagrangian of Motion	29
		2.2.2	Extended Hamilton Principle	31
		2.2.3	Order-Three Equations of Motion	33
3	**Parametric System Identification**			**37**
	3.1	Theoretical Modeling		37
		3.1.1	Equation of Motion	37
		3.1.2	Single-Mode Response	38
		3.1.3	Frequency-Response and Force-Response Equations	41
	3.2	Experimental Procedure		42
		3.2.1	Linear Natural Frequencies	43
		3.2.2	Determination of the Beam Displacement	43
	3.3	Parameter Estimation Procedure		45
		3.3.1	Estimation of the Damping Coefficients	46
		3.3.2	Nonlinearity Estimation	47
		3.3.3	Curve-Fitting the Frequency-Response Data	47
		3.3.4	Fixing f and ω_n	48
		3.3.5	Critical Forcing Amplitude	50
	3.4	Results		51

	3.4.1	Third-Mode Estimation Results	51
	3.4.2	Comparison with Curve-Fitting Method	55
3.5	Closure		58

4 Determination of Jump Frequencies   60

- 4.1 Theory 61
  - 4.1.1 Frequency-Response Function 61
  - 4.1.2 Sylvester Resultant 63
  - 4.1.3 Critical Forcing Amplitude 64
  - 4.1.4 Jump Frequencies 65
  - 4.1.5 Gröbner Basis 65
- 4.2 Results 67
- 4.3 Closure 68

5 Energy Transfer Between Widely Spaced Modes Via Modulation   69

- 5.1 Planar Motion 70
  - 5.1.1 Test Setup 70
  - 5.1.2 Experimental Results 73
  - 5.1.3 Reduced-Order Model 79
  - 5.1.4 Numerical Results 84
- 5.2 Nonplanar Motion 86
  - 5.2.1 Experiments with a Circular Rod 86

		5.2.2	Reduced-Order Model	92

      5.2.3   Numerical Results 94

  5.3 Closure 96

6 Experiments with a Cantilever Plate   99

  6.1 Test Setup 100

  6.2 Results 101

      6.2.1   RUN I: External Combination and Two-to-One Internal Resonances 102

      6.2.2   RUN II: Two-to-One and Zero-to-One Internal Resonances 103

      6.2.3   RUN III: Quasiperiodic Motion and Three-to-One Internal Resonance 107

  6.3 Closure 110

7 Conclusions and Recommendation for Future Work   113

  7.1 Summary 113

  7.2 Suggestions for Future Work 115

List of Figures

2.1 A schematic of a vertically mounted metallic cantilever beam undergoing flexural-flexural-torsional motions. 23

2.2 3-2-1 Euler-angle rotations. 24

2.3 Deformation of a beam element along the neutral axis. 26

2.4 Initial and deformed positions of an arbitrary point P. 27

3.1 Experimentally and theoretically obtained third-mode frequency-response curves for $a_b = 0.2g$ and $\omega_3 = 49.094$ Hz using the linear damping model. 48

3.2 Experimentally and theoretically obtained third-mode frequency-response curves for $a_b = 0.23g$ and $\omega_3 = 49.06$ Hz using the linear damping model. 49

3.3 Experimentally and theoretically obtained third-mode frequency-response curves for $a_b = 0.1g$, $0.15g$, and $0.2g$ using the linear damping model. 53

3.4 Experimentally and theoretically obtained third-mode frequency-response curves for $a_b = 0.1g$, $0.15g$, and $0.2g$ using the nonlinear damping model. 53

3.5 Experimentally and theoretically obtained third-mode force-response curves using the linear and nonlinear damping models for $\Omega = 48.891$ Hz. 54

3.6 Overshoot in the peak of the third-mode frequency-response curve obtained using the linear damping model. 55

3.7 Comparison of the fourth-mode frequency-response curves obtained using the proposed technique and the curve-fitting method for $a_b = 0.075g$ and $0.1g$ using the linear damping model. 56

3.8 Comparison of the fourth-mode frequency-response curves obtained using the proposed technique and the curve-fitting method for $a_b = 0.075g$ and $0.1g$ using the nonlinear damping model. Note: In the legend, NL stands for Nonlinear. 57

3.9 Comparison of the fourth-mode force-response curves obtained using the proposed technique and the curve-fitting method for $\Omega = 95.844$ Hz using the linear and nonlinear damping models. 57

4.1 Typical frequency-response curves of a Duffing oscillator with (a) softening nonlinearity and (b) hardening nonlinearity. Dashed lines (- -) indicate unstable solutions and SN refers to a saddle-node bifurcation. 61

4.2 Frequency-response curves obtained using (a) $f = f_{cr}$ and (b) $f = 8.82$. The asterisk in (a) indicates the inflection point and the circles in (b) indicate the jump-up and jump-down points. 67

5.1 Experimental setup. 71

5.2 Frequency-response curve of the third mode when $a_b = 0.8g$. 74

5.3 Input and response time traces at $\Omega = 16.547$ Hz when $a_b = 0.8g$. 75

5.4 Input and response FFTs at $\Omega = 16.547$ Hz when $a_b = 0.8g$. 76

5.5 Time traces and FFT of the chaotic motion observed at $\Omega = 16.531$ Hz when $a_b = 0.8g$. 77

5.6 Force-response curve of the third mode when $\Omega = 16$ Hz. 78

5.7 Input and response time traces at $\Omega = 17.109$ Hz when $a_b = 2.3g$. 79

5.8 Input and response FFTs at $\Omega = 17.109$ Hz when $a_b = 2.3g$. 80

5.9 Input and response time traces at $\Omega = 17.547$ Hz when $a_b = 2.97g$. 81

5.10 Input and response FFTs at $\Omega = 17.547$ Hz when $a_b = 2.97g$. 82

5.11 Displacement time trace and FFT at $\Omega = 0.977$ when $a_b = 1.5g$. 85

5.12 Displacement time trace and FFT at $\Omega = 0.945$ when $a_b = 1.5g$. 86

5.13 Displacement FFTs at (a) $\Omega = 0.984$ when $a_b = 1g$, (b) $\Omega = 0.972$ when $a_b = 2g$, and (c) $\Omega = 0.9678$ when $a_b = 2.5g$. 87

5.14 Frequency-response curves of the fifth mode of a circular rod for an excitation amplitude of $2g$ rms (S. Nayfeh and Nayfeh, 1994). 88

5.15 A short portion of the long time history of a typically weakly modulated motion of a circular rod (S. Nayfeh and Nayfeh, 1994). 89

5.16 Time traces of a typically weakly modulated motion of a circular rod (S. Nayfeh and Nayfeh, 1994). 89

5.17 Power spectrum of a typically weakly modulated motion of a circular rod (S. Nayfeh and Nayfeh, 1994). 90

5.18 A short portion of the long time history of a typically strongly modulated motion of a circular rod (S. Nayfeh and Nayfeh, 1994). 91

5.19 Time traces of a typically strongly modulated motion of a circular rod (S. Nayfeh and Nayfeh, 1994). 91

5.20 Power spectrum of a typically strongly modulated motion of a circular rod (S. Nayfeh and Nayfeh, 1994). 92

5.21 Time traces of in-plane and out-of-plane motion of beam at $\Omega = 82.75$ Hz. 95

5.22 FFTs of in-plane and out-of-plane motion of beam at $\Omega = 82.75$ Hz. 96

5.23 Time traces of in-plane and out-of-plane motion of beam at $\Omega = 82.59$ Hz. 97

5.24 FFTs of in-plane and out-of-plane motion of beam at $\Omega = 82.59$ Hz. 98

6.1 Experimental setup. 100

6.2 Input and response FFTs for $a_b = 2.7g$ and $\Omega = 316.81$ Hz. 103

6.3 Response time trace for $a_b = 2.7g$ and $\Omega = 316.81$ Hz. 104

6.4 Poincaré section showing two-period quasiperiodic motion. 104

6.5 Input and response FFTs for $a_b = 2.7g$ and $\Omega = 315$ Hz. 105

6.6 Response time trace for $a_b = 2.7g$ and $\Omega = 315$ Hz. 106

6.7 Response FFT for $a_b = 4.5g$ and $\Omega = 311$ Hz. 106

6.8 Pseudo-phase plane trajectory showing two-to-one internal resonance. 107

6.9 Response FFT for $a_b = 4.5g$ and $\Omega = 304.5$ Hz. 107

6.10 Response FFT for $a_b = 3g$ and $\Omega = 300.94$ Hz. 108

6.11 Response FFT for $a_b = 2g$ and $\Omega = 109.41$ Hz. 109

6.12 Response FFT for $a_b = 2g$ and $\Omega = 109.39$ Hz. 109

6.13 Response FFT for $a_b = 2g$ and $\Omega = 109.35$ Hz. 110

6.14 Input and response FFTs for $a_b = 2g$ and $\Omega = 109.35$ Hz. 111

6.15 Response FFT for $a_b = 2g$ and $\Omega = 108.4$ Hz. 112

6.16 Response FFT for $a_b = 2g$ and $\Omega = 108.5$ Hz. 112

List of Tables

3.1 Experimentally determined third-mode natural frequency. 43

3.2 Some constants and their values. 45

3.3 Coordinates of the peak of the frequency-response curve. 46

3.4 Modified values of ω_3. The superscripts l and n refer to the linear and nonlinear damping models, respectively. 50

3.5 Estimated values of the third-mode viscous damping factor ζ of the linear damping model. 52

3.6 Estimated values of the third-mode damping coefficients ζ and $\bar{c}$ of the nonlinear damping model. 52

3.7 Estimated values of the third-mode effective nonlinearity α. The superscripts l and n refer to the linear and nonlinear damping models, respectively. 52

3.8 Comparison of the estimates of the damping factor ζ and the effective nonlinearity α for the fourth mode using the linear damping model. The superscripts p and cf refer to the proposed estimation technique and the curve-fitting method, respectively. 56

3.9 Comparison of the estimates of the fourth-mode damping coefficients ζ and $\bar{c}$ and the effective nonlinearity α using the nonlinear damping model. The superscripts p and cf refer to the proposed estimation technique and the curve-fitting method, respectively. . 56

5.1 The first six in-plane natural frequencies – experimental and analytical values. 72

Chapter 1

Introduction

1.1 Motivation

The beam is one of the fundamental elements of an engineering structure. It finds use in varied structural applications. Moreover, structures like helicopter rotor blades, spacecraft antennae, flexible satellites, airplane wings, gun barrels, robot arms, high-rise buildings, long-span bridges, and subsystems of more complex structures can be modeled as a beam-like slender member. Therefore, studying the static and dynamic response, both theoretically and experimentally, of this simple structural component under various loading conditions would help in understanding and explaining the behavior of more complex, real structures under similar loading.

Interesting physical phenomena occur in structures in the presence of nonlinearities, which cannot be explained by linear models. These phenomena include jumps, saturation, subharmonic, superharmonic, and combination resonances, self-excited oscillations, modal interactions, and chaos. In reality, no physical system is strictly linear and hence linear models of physical systems have limitations of their own. In general, linear models are applicable only in a very restrictive domain like when the vibration amplitude is very small. Thus, to accurately identify and understand the dynamic behavior of a structural system under general loading conditions, it is essential that nonlinearities present in the system also be modeled and studied.

In continuous (or distributed-parameter) systems like structures, nonlinearities essentially couple the linearly uncoupled normal modes, and this coupling could lead to modal interactions (i.e., interaction between the modes), resulting in the transfer of energy among modes. Experiments have demonstrated that sometimes energy is transferred from a directly excited high-frequency mode to a low-frequency mode, which may be extremely dangerous because the response amplitude of the low-frequency mode can be very large compared with that of the directly excited high-frequency mode. A lot of research is under way to understand this and other interesting nonlinear phenomena.

In this dissertation, we study both experimentally and theoretically the nonlinear vibrations of two flexible, metallic cantilever beams under transverse (or external or additive) harmonic excitations. In particular, we investigate the transfer of energy between modes whose natural frequencies are widely spaced – in the absence and presence of an internal resonance. We also develop an experimental parametric identification technique to estimate the linear and nonlinear damping coefficients of a beam along with its effective nonlinearity. In addition, we study experimentally the response of a rectangular, metallic cantilever plate under transverse harmonic excitation.

1.2 Types of Nonlinearity

In theory, nonlinearity exists in a system whenever there are products of dependent variables and their derivatives in the equations of motion, boundary conditions, and/or constitutive laws, and whenever there are any sort of discontinuities or jumps in the system. Evan-Iwanowski (1976), Nayfeh and Mook (1979), and Moon (1987) explain the various types of nonlinearities in detail along with examples. Here, we briefly describe the relevant nonlinearities. In structural mechanics, nonlinearities can be broadly classified into the following categories:

1. *Damping* is essentially a nonlinear phenomenon. Linear viscous damping is an idealization. Coulomb friction, aerodynamic drag, hysteretic damping, etc. are examples of nonlinear damping.

2. *Geometric* nonlinearity exists in systems undergoing large deformations or deflections. This nonlinearity arises from the potential energy of the system. In structures, large deformations usually result in nonlinear strain- and curvature-displacement relations. This type of nonlinearity is present, for example, in the equation governing the large-angle motion of a simple pendulum, in the nonlinear strain-displacement relations due to mid-plane stretching in strings, and due to nonlinear curvature in cantilever beams.

3. *Inertia* nonlinearity derives from nonlinear terms containing velocities and/or accelerations in the equations of motion. It should be noted that nonlinear damping, which has similar terms, is different from nonlinear inertia. The kinetic energy of the system is the source of inertia nonlinearities. Examples include centripetal and Coriolis acceleration terms. It is also present in the equations describing the motion of an elastic pendulum (a mass attached to a spring) and those describing the transverse motion of an inextensional cantilever beam.

4. When the constitutive law relating the stresses and strains is nonlinear, we have the so-called *material* nonlinearity. Rubber is the classic example. Also, for metals, the nonlinear Ramberg-Osgood material model is used at elevated temperatures.

5. Nonlinearities can also appear in the *boundary conditions*. A nonlinear boundary condition exists, for instance, in the case of a pinned-free rod attached to a nonlinear torsional spring at the pinned end.

6. Many other types of nonlinearities exist: like the ones in systems with impacts, with backlash or play in their joints, etc.

It is interesting to note that the majority of physical systems belong to the class of weakly nonlinear (or quasi-linear) system. For certain phenomena, these systems exhibit a behavior only slightly different from that of their linear counterpart. In addition, they also exhibit phenomena which do not exist in the linear domain. Therefore, for weakly nonlinear structures, the usual starting point is still the identification of the linear natural frequencies and mode shapes. Then, in the analysis, the dynamic response is usually described in terms of its linear natural frequencies and mode shapes. The effect of the small nonlinearities is seen in the equations governing the amplitude and phase of the structure response.

1.3 Literature Review

The sheer quantity of material published in the field of nonlinear vibrations of beams makes it almost impossible to list all of them. But the necessary and relevant articles and books will be included here to give a gist of the research done in this area. Unfortunately, the review is restricted only to the English literature.

1.3.1 Beam Theories

A very detailed and interesting historical account of the development of the theory of elasticity, including the beam bending problem, is given by Love (1944) and Timoshenko (1983). Beginning with the works of Galileo, they describe the refinements made to the beam theory by the Bernoullis, Euler, Coulomb, Saint-Venant, Poisson, Kirchhoff, Rayleigh, and Timoshenko, to name just a few. The present day beam theories still use the same basic principles developed decades, and in some cases centuries, ago. In the literature, the words *bar* and *rod* are also used for a beam; and beams with cross-sectional areas having approximately equal principal moments of inertia are referred to as *compact* beams.

The popular beam theories in use today are (a) the exact elasticity equations, (b) the Euler-Bernoulli beam theory, and (c) the Timoshenko beam theory. The theory of elasticity approach has a major drawback that only a few problems can be solved exactly (Cowper, 1968), and hence it is not very attractive. The Euler-Bernoulli beam theory (Shames and Dym, 1985) assumes that plane cross sections, normal to the neutral axis before deformation, continue to remain plane and normal to the neutral axis and do not undergo any strain in their planes (i.e., their shape remains intact). In other words, the warping and transverse shear-deformation effects and transverse normal strains are considered to be negligible and hence ignored. These assumptions are valid for slender beams. The no-transverse-shear assumption means that the rotation of the cross sections is due to bending alone. But for problems where the beam is thick, or high-frequency modes are excited, or the beam is made up of a composite material, the transverse shear is not negligible. Incorporating the effect of transverse shear deformation into the Euler-Bernoulli beam model gives us the Timoshenko beam theory (Timoshenko, 1921,1922; Meirovitch, 1967; Shames and Dym, 1985). In this theory, to simplify the derivation of the equations of motion, the shear strain is assumed to be uniform over a given cross section. In turn, a shear correction factor is introduced to account for this simplification, and its value depends on the shape of the cross section (Timoshenko, 1921; Cowper, 1966,1968). In the presence of transverse shear, the rotation of the cross section is due to both bending and transverse (or out-of-plane) shear deformation.

A linear beam model would suffice when dealing with small deformations. But when the deformations are moderately large, for accurate modeling, several nonlinearities also need to be included. It is impossible to come up with a very general three-dimensional beam theory incorporating all possible nonlinearities and secondary effects, like rotatory inertia, shear deformation, warping, damping, static

Chapter 1. Introduction   5

deformation, etc. Usually insignificant nonlinearities and secondary effects are dropped to (a) simplify various expressions, (b) make the model manageable, and (c) facilitate solving the model equations. The selection of nonlinearities and secondary effects to be dropped depends on the beam properties (dimensions, material, etc.) and configuration (loading and boundary conditions).

Most of the nonlinear theories of transverse beam vibrations deal with the effect of midplane stretching for the case of a simply supported uniform beam with an infinite axial restraint. Woinowsky-Krieger (1950) and Burgreen (1951) considered free oscillations of a beam having hinged ends a fixed distance apart. Their equation of motion contained a nonlinear term due to midplane stretching, which results in nonlinear strain-displacement relations. They gave the solution in terms of elliptic functions and also found that the frequency of vibration varies with the amplitude. Burgreen also studied, both theoretically and experimentally, the effects of a compressive axial load. Eisley (1964) analyzed the effect of an axial periodic load on the motion of a hinged beam. He also studied the stability of the periodic beam response. The above theories are in good agreement with the experiments of Ray and Bert (1969). Evensen (1968) analyzed the effect of midplane stretching on the vibrations of a uniform beam with immovable ends for simply supported, clamped, and clamped-simply supported cases. Busby and Weingarten (1972) used the finite-element method to formulate the nonlinear differential equations of a beam under periodic loading. Both simply supported and clamped boundary conditions were considered. Ho, Scott, and Eisley (1975,1976) accounted for the midplane stretching in the study of large-amplitude nonplanar whirling motions of a simply supported beam.

Bolotin (1964) showed that, for beams, inertia nonlinearity effects are more significant than geometric nonlinearity effects. Atluri (1973) studied the nonlinear vibrations of a hinged beam with one end free to move in the axial direction. He included rotatory inertia and nonlinearities due to inertia and geometry, but ignored the effects of midplane stretching and transverse shear deformation. Using the method of multiple scales to solve the governing equations, he found out that the effective nonlinearity depends on the contributions of the geometric and inertia nonlinearity terms, which in turn vary with the mode number. He also noted that the inertia nonlinearity is of the softening type. Crespo da Silva and Glynn (1978a,b) systematically derived the nonlinear equations of motion and boundary conditions governing the flexural-flexural-torsional motion of isotropic, inextensional beams. They included nonlinearities due to inertia and geometry up to order three and showed that the nonlinearities arising from the curvature (geometry) are of the same order of magnitude as those due to inertia. Using the equations derived by Crespo da Silva and Glynn (1978a,b), Pai and Nayfeh (1990b) and

Anderson, Nayfeh, and Balachandran (1996b) investigated the nonlinear motions of cantilever beams and observed that, for the first mode, the geometric nonlinearity, which is of the hardening type, is dominant; whereas for the second and higher modes, the inertia nonlinearity, which is of the softening type, becomes dominant.

Nordgren (1974) developed a computational method for finite-amplitude three-dimensional motions of inextensible beams and successfully used it for problems encountered in offshore pipe laying operations. Epstein and Murray (1976) formulated a theory for the three-dimensional large deformation analysis of thin-walled beams of arbitrary open cross section. Numerical solutions for the post-buckling behavior of "I" beams obtained using this theory compared very well with experimental data. Hodges and Dowell (1974) developed nonlinear equations of motion with quadratic nonlinearities to describe the dynamics of slender, rotating, extensional helicopter rotor blades undergoing moderately large deformations. Dowell, Traybar, and Hodges (1977) experimentally studied the large deformation of a simple, non-rotating cantilever beam under a gravity tip load to evaluate the theory of Hodges and Dowell (1974). Agreement was reasonably good for small bending deflections, but systematic differences occured for larger deflections. Rosen and Friedmann (1979) derived a more accurate set of equations than those of Hodges and Dowell (1974) by including some nonlinear terms of order three. Their numerical results are in good agreement with the experimental data obtained by Dowell, Traybar, and Hodges (1977). Rosen, Loewy, and Mathew (1987a,b) derived equations for analyzing the nonlinear coupled bending-torsion motions of pretwisted rods. Comparison of the static results with those from experiments is very good. Rosen, Loewy, and Mathew (1987c) extended the above study to the dynamic case and once again obtained very good agreement with the experimental results. Danielson and Hodges (1987) and Hodges (1987b) used the concept of local rotation to account for the warping effects and obtained a simple matrix expression for the strain components of a beam. Kane, Ryan, and Banerjee (1987) developed a comprehensive theory dealing with small vibrations of a beam attached to a base that is performing an arbitrary but prescribed three-dimensional motion (translation and rotation). This theory is applicable to beams with arbitrary cross section and spatially varying material properties. Through an example, they highlighted the deficiencies in popular multibody dynamic-simulation computer programs. Hinnant and Hodges (1988) developed a program, which combines multibody and finite-element technology, to study the nonlinear static and linearized dynamic behavior of structures. The results of this program match very closely the experimental data obtained by Dowell, Traybar, and Hodges (1977). Crespo da Silva, Zaretzky, and Hodges (1991) studied the static equilibrium deflection and natural frequencies associated with infinitesimally small oscillations about the static equilibrium and obtained results almost identical with the finite-element results of Hinnant and Hodges (1988) and the experimental data of Dowell, Traybar, and Hodges (1977).

Crespo da Silva and Hodges (1986a,b) formulated the nonlinear differential equations of motion for a rotating beam, with the objective of retaining all possible contributions due to cubic nonlinearities, and investigated the influence of these terms on the motion of a helicopter rotor blade. They found out that the most significant cubic nonlinear terms are those associated with the structural geometric nonlinearity in the torsion equation. Equations describing the nonlinear flexural-flexural-torsional-extensional dynamics of beams were formulated by Crespo da Silva (1988a,1991). Nonlinearities due to curvature, inertia, and extension were accounted for in a mathematically consistent manner. Pai and Nayfeh (1990a) also developed the nonlinear equations describing the extensional-flexural-flexural-torsional vibrations of slewing or rotating metallic and composite beams. The equations contain structural coupling terms and quadratic and cubic nonlinearities due to curvature and inertia. Pai and Nayfeh (1992) extended the above model to include the effect of transverse shear deformation. Pai and Nayfeh (1994) derived a geometrically exact nonlinear beam model for naturally curved and twisted solid composite rotor blades undergoing large vibrations, accounting for warpings and three-dimensional stress effects.

While deriving the equations of motion describing the three-dimensional large-amplitude motion of a beam, three successive Euler-angle-like rotations are used to relate the deformed and undeformed configurations. Hodges (1987a) reviewed and compared the standard ways of representing finite rotation in rigid-body kinematics, including orientation angles, Euler parameters, and Rodrigues parameters. Hodges, Crespo da Silva, and Peters (1988) discussed some of the common mistakes in the nonlinear modeling of a cantilever beam.

Strutt (1945), well known as Lord Rayleigh, was the first to consider the effect of rotatory inertia in his book 'The Theory of Sound', which first appeared in 1877. This was later extended by Timoshenko (1921,1922) to include the effect of transverse-shear deformation. Timoshenko (1921) showed, for a simply supported beam, that the correction for shear is four times greater than the correction for rotatory inertia and that the shear and rotatory inertia effects increase with an increase in the mode number. Mindlin (1951) showed that the rotatory inertia effect is almost invariably small for lower modes of plates. Huang (1961) studied the influence of rotatory inertia and shear deformation on the natural frequencies and mode shapes of uniform Timoshenko beams with simple boundary conditions. He showed that the influence of the two secondary effects on the natural frequencies increases with an increase in the mode number or the cross-section dimensions. But the comparative influence on the normal mode shapes seems to be very small. Murty (1970) derived linear approximate equations for transverse vibrations of uniform short beams including shear deformation and rotatory inertia effects. His values of the natural frequencies were in better agreement with the experimental trends compared to those obtained using the shear correction factors suggested by Timoshenko (1921) and Cowper (1966). Adams and Bacon (1973) stated that the shear-deformation effect is a function of the aspect ratio (i.e., ratio of length to thickness) and is less than 1% for isotropic materials with aspect ratio greater than twenty.

Nayfeh (1973a) studied the nonlinear transverse vibration of inhomogeneous beams with finite axial restraints, taking into account the effects of transverse shear and rotatory inertia. The results show that the frequency of vibration increases with amplitude and axial restraint and that the transverse shear and rotatory inertia decrease the natural frequency. Rao, Raju, and Raju (1976) studied large-amplitude free vibrations of beams including the shear-deformation and rotatory-inertia effects. Using their nonlinear beam model, they showed that the two secondary effects have negligible influence when $l/r > 100$, where l and r denote the beam length and radius of gyration, respectively. Sinclair (1979), using his nonlinear beam model, concluded that the effects of shear deformation and longitudinal deformation (i.e., beam extension) are of the same order. Crespo da Silva (1988b,1991) showed that beams with one end free to move behave essentially as if they were inextensional when the value of EAl^2/D_η or EAl^2/D_ζ (i.e., the ratio l/r squared) is large, where E, A, D_η, and D_ζ denote Young's modulus, area of cross section, and flexural rigidities, respectively. In most studies with slender beams, the out-of-plane (transverse) shear induced warping is usually neglected, but the torsion induced warping is used to account for its influence on the torsional rigidity and hence the torsional frequencies (Timoshenko and Goodier, 1970; Rosen and Friedmann, 1979; Crespo da Silva, 1988a). In the case of slender beams, Poisson's effect is generally very small and hence is also neglected.

Caughey and O'Kelly (1961) studied the effect of weak damping on the natural frequencies of a multi-degree-of-freedom linear system. They showed that the highest natural frequency is always decreased by damping, but the lower natural frequencies may either increase or decrease, depending on the form of the damping matrix. Adams and Bacon (1973) showed experimentally that air damping in beams is significant, and that it is a function of the beam geometry, mode shape, amplitude, and frequency of vibration. Anderson, Nayfeh, and Balachandran (1996b), while studying the nonlinear vibrations of a parametrically excited beam, showed that inclusion of quadratic air damping in the analytical model significantly improves the agreement between experimental and theoretical results. In the case of harmonic excitation, the energy loss for both viscous damping and structural damping is proportional to the square of the displacement amplitude. Thus, for structurally damped systems subjected to a harmonic excitation, it is convenient to replace the structural damping by an equivalent viscous damping term (Meirovitch, 1997).

Evensen (1968) showed that, for higher modes of vibration, the amplitude-frequency curves for a clamped-clamped beam or a clamped-supported beam tend to approach that of a simply supported beam. In other words, the influence of the boundary conditions on the response becomes less and less pronounced as the mode number increases. Aravamudan and Murthy (1973) also observed the same behavior while studying the effect of time-dependent boundary conditions on the nonlinear vibrations of beams. In reality, an ideal clamped boundary condition, for example, is impossible to obtain. Thus, to model real joints, it becomes necessary to add damping and mass elements in addition to rotational and translational springs (Gorman, 1975). Tabaddor (2000) replaced the clamped-end boundary condition of a cantilever beam by a torsional spring possessing linear and cubic stiffness components. This helped improve dramatically the agreement between experimental and theoretical results. Chun (1972) derived expressions for the mode shapes and natural frequencies of a beam hinged at one end by a rotational spring and the other end free. Arafat and Nayfeh (2001) studied the influence of nonlinear boundary conditions on the nonplanar autoparametric responses of an inextensible cantilever beam, whose free end was restrained by nonlinear springs. They found out that the effective nonlinearity is sensitive to the stiffness components of the springs.

It is well known that static deflection of a nonlinear beam affects its natural frequencies. Governing equations, originally containing only cubic nonlinearities, would also have quadratic nonlinearities when a static deflection is present. Sato, Saito, and Otomi (1978) studied the influence of gravity on the parametric resonance of a simply supported horizontal beam carrying a concentrated mass at an arbitrary point. Their results show that the change in the value of the first natural frequency is proportional to the static deflection caused by the concentrated mass and that the static deflection has a softening effect, which depends on the location and weight of the concentrated mass. This softening effect could overcome the hardening terms if the static deflection is large or when the beam is very slender (Hughes and Bert, 1990).

1.3.3 Modal Interactions

In recent years, many examples of modal interactions have been studied both experimentally and analytically. Modal interactions may be the result of internal (autoparametric) resonances, external combination resonances, parametric combination resonances, or nonresonant interactions (Nayfeh and Mook, 1979; Nayfeh, 2000). Internal resonances may occur in systems where the linear natural frequencies ω_i are commensurate or nearly commensurate; that is, there exist non-zero integers k_i such that $k_1\omega_1 + k_2\omega_2 + \cdots + k_n\omega_n \approx 0$. When the nonlinearity is cubic, internal resonances can occur if $\omega_n \approx \omega_m$, $\omega_n \approx 3\omega_m$, $\omega_n \approx |2\omega_m \pm \omega_k|$, or $\omega_n \approx |\omega_m \pm \omega_k \pm \omega_l|$. When the nonlinearity is quadratic, besides the above resonances, internal resonances can also occur if $\omega_n \approx 2\omega_m$ or $\omega_n \approx \omega_m \pm \omega_k$. External combination resonances may occur if the excitation frequency Ω is commensurate or nearly commensurate with two or more natural frequencies. For systems with cubic nonlinearities, external combination resonance may occur if $\Omega \approx |2\omega_m \pm \omega_n|$, $\Omega \approx \frac{1}{2}(\omega_m \pm \omega_n)$, or $\Omega \approx |\omega_m \pm \omega_n \pm \omega_k|$. If quadratic nonlinearities are added, additional external combination resonances may occur if $\Omega \approx \omega_m \pm \omega_n$. Parametric combination resonances may occur whenever $\Omega \approx \omega_m \pm \omega_n$. For a detailed account of many combination resonances in different mechanical and structural systems, we refer the reader to Evan-Iwanowski (1976).

In contrast, nonresonant modal interactions channel energy from a high-frequency mode to a low-frequency mode even if there is no special relationship between their frequencies. The only requirement for such an energy transfer is that the modes be widely spaced; that is, $\omega_i \gg \omega_j$. The signature of this type of modal interaction appears to be the presence of asymmetric sidebands around the high-frequency component in the response spectrum, with the sideband spacing being approximately equal to the natural frequency of the low-frequency mode. The sidebands and their asymmetry point to phase and amplitude modulations of the high-frequency mode. This type of interaction, where energy is transferred from a high-frequency to a low-frequency mode via modulation, is sometimes also referred to as zero-to-one internal resonance or as Nayfeh's resonance (Langford, 2001).

Resonant Modal Interactions

McDonald (1955) worked with the governing equations developed by Woinowsky-Krieger (1950) and Burgreen (1951), but did not consider axial prestressing. He represented the beam response in terms of the linear mode shapes and solved the nonlinear equations for the coefficients in terms of elliptic functions. He concluded that the problem is inherently nonlinear even for small-amplitude vibrations, there is dynamic coupling between the modes, and the frequencies of the various modes are functionally related to the amplitudes of all of the modes. Henry and Tobias (1961) studied theoretically and experimentally an undamped two-degree-of-freedom system when the two natural frequencies are almost equal. They discussed the conditions necessary for the existence of motion in a single mode and for the mode at rest to lose stability. Ginsberg (1972) examined the forced response of a two-degree-of-freedom system with equal frequencies. Two-mode responses were observed, which disappeared when the damping was increased beyond a critical value.

Nayfeh, Mook, and Sridhar (1974) used the method of multiple scales to obtain the nonlinear response of structural elements subjected to harmonic excitation, with a special emphasis on modal interactions. In the case of a clamped-hinged beam with the ratio of their first two natural frequencies being close to 1/3, they observed that it is possible for the response to be dominated by the first mode when the excitation frequency is near the second natural frequency. To study the stability of the periodic motions, they perturbed the amplitudes and phases of the directly- and indirectly-excited modes, linearized the modulation equations describing the evolution of the amplitudes and phases of the excited modes, and obtained a set of first-order equations with constant coefficients governing the small disturbances. But in earlier stability studies, the periodic solutions were perturbed and were put back into the nonlinear equations of motion; thus resulting in coupled equations of the Mathieu type, which would require in general more effort to determine the stability. Nayfeh, Mook, and Lobitz (1974) extended the above work to structural elements having complicated boundaries and/or composition. Tezak, Mook, and Nayfeh (1978) studied the nonlinear response of a hinged-clamped beam when the excitation frequencies are away from the natural frequencies, but near a multiple or combination of the natural frequencies. They observed multiple jumps in the response curves and the excitation of two modes, initially at rest, due to a combination resonance.

Nonplanar motions are possible when the frequencies of an in-plane and an out-of-plane mode are involved in an internal resonance. Murthy and Ramakrishna (1965) studied theoretically and experimentally the nonplanar motion near resonance of stretched strings. They observed that beyond a critical forcing value, nonplanar whirling (or ballooning) motions exist for a range of frequency values. Miles (1965) investigated in detail the stability of such motions in the absence of damping. Anand (1966,1969) studied the nonlinear motions of stretched strings with the addition of viscous damping and determined their stabilities.

Haight and King (1971,1972) theoretically and experimentally investigated the responses of compact cantilever beams to external (additive) and parametric (multiplicative) excitations. Nonlinear inertia with linear curvature was considered while deriving the equations of motion, which were later solved using the Galerkin method. They found out that, for certain values of excitation amplitudes and frequencies, the planar response is unstable, and a nonplanar motion gets parametrically excited. But they did not quantify the nonplanar motions. Their results show that, when the planar motion loses stability, every point on the cross section traces an elliptical path in a plane normal to the rod axis and that the planar instability is not a large-amplitude phenomenon.

Ho, Scott, and Eisley (1975) analyzed the forced response of a simply supported, compact beam. They found out that, as the beam approached resonance, it would start whirling. They also determined the in-plane and out-of-plane responses and the stability zones of such motions. Crespo da Silva and Glynn (1978b) studied the nonlinear response of a compact cantilever beam under external excitation, using the equations derived by the same authors (1978a). They obtained response curves similar to those of Haight and King (1972). This work was extended by Crespo da Silva and Glynn (1979a,b) to clamped-clamped/sliding beams and to fixed-free beams with support asymmetry. Crespo da Silva (1978a) determined the nonlinear response of a column with a follower force (Beck's column) subjected to either a distributed periodic lateral excitation or a support excitation. Crespo da Silva (1978b) extended the above problem to nonplanar motions by considering an internal resonance. Crespo da Silva and Zaretzky (1990) studied the nonlinear responses of compact cantilever and clamped-pinned/sliding beams in the presence of a one-to-three internal resonance. Zaretzky and Crespo da Silva (1994a) experimentally investigated the nonlinear modal coupling in the response of compact cantilever beams and obtained excellent agreement with the theoretical predictions of Crespo da Silva and Glynn (1978b).

Hyer (1979) used the equations of Haight and King (1972) and studied the whirling motions of an undamped cantilever beam, with square or circular cross section, under external excitation. He concluded that stable whirling motions exist over a range of frequency near the resonant frequencies of the beam and that no unstable whirling motions are present in this range. Crespo da Silva (1980) pointed out that nonplanar whirling motions can indeed be unstable even when damping and nonlinear curvature are not considered. Pai and Nayfeh (1990b) analyzed the nonlinear nonplanar oscillations of a cantilever beam under external excitations using the equations developed by Crespo da Silva and Glynn (1978a,b). They obtained quantitative results for nonplanar motions and investigated their dynamic behavior. They found the nonplanar motions to be either steady whirling motions, whirling motions of the beating type (quasiperiodic motion), or chaotic.

Simplifying the equations, derived by Crespo da Silva (1988a), for beams with high torsional frequencies and neglecting rotatory inertia, Crespo da Silva (1988b) investigated the planar response of an extensional beam to a periodic excitation. The results show that the effect of the nonlinearity due to midplane stretching is dominant and that neglecting the nonlinearities due to curvature and inertia does not introduce significant error in the results. Also, unlike the response of an inextensional beam, the single-mode response of an extensional beam is always hardening.

When the torsional frequencies of the beam are much higher than its bending frequencies, the torsional inertia has no significant effect on the beam motion. In such a case, the torsional deformation is only due to the nonlinear coupling between in-plane and out-of-plane bending. But for beams having, for example, a cross section with high aspect ratio, the first torsional natural frequency is of the order of a lower bending natural frequency. Then, the nonlinear coupling between torsional and bending motions may cause an exchange of energy between such motions. In the governing equations of inextensional beams with high torsional frequencies, only cubic nonlinearities are present. But when torsional dynamics is accounted for as in the case of beams with low torsional frequencies, nonlinear quadratic terms also appear in the equations of motion. Crespo da Silva and Zaretzky (1994) examined such coupling in inextensional beams by taking into account the torsional dynamics of the beam. Considering a one-to-one internal resonance between an in-plane bending mode and a torsional mode and exciting the in-plane mode, they observed that coupled bending-torsion exists. Also, within certain regions of the excitation amplitude, the in-plane bending component of the coupled response saturates so that any further energy pumped into the system is transferred to the torsional motion via the internal resonance. Zaretzky and Crespo da Silva (1994b) extended the work of Crespo da Silva and Zaretzky (1994) to the case of an internal combination resonance involving modes associated with bending in two directions and torsion. Their results show that the occurence of a coupled bending-torsion response depends on the physical properties of the beam. When the coupled response exists, the out-of-plane bending and torsional components are simply related by a constant.

Tso (1968) studied parametrically induced torsional vibrations in a cantilever beam, of rectangular cross section, under dynamic axial loading. He found out that, when the applied frequency is close to twice one of the torsional natural frequencies, the corresponding torsional modes may be excited parametrically. In addition, the frequency range, over which torsional vibrations are present, increases tremendously when the applied frequency is near a longitudinal natural frequency close to twice one of the torsional frequencies. Dugundji and Mukhopadhyay (1973) investigated the response of a cantilever beam excited parametrically at a frequency close to the sum of the first bending and torsional frequencies. They observed a large contribution from the low-frequency first bending mode besides those at the excitation frequency and the first torsional mode. Dokumaci (1978) studied, both theoretically and experimentally, the coupled bending-torsion vibrations, due to combination resonances, of a cantilever beam under lateral parametric excitation. He obtained the instability boundaries with and without damping and found out that damping widens the unstable regions. His experimental and theoretical results match very well.

Shyu, Mook, and Plaut (1993a,b) studied the nonlinear response of a slender cantilever beam subjected to primary- and secondary-resonance excitations, including the effects of static deflection. Shyu, Mook, and Plaut (1993c) extended the above study to nonstationary excitations, where the excitation frequency is varied with time. They found out that, when the sweep rate (i.e., rate of change of excitation frequency) is small, the nonstationary amplitude closely follows the stationary amplitude; otherwise, there is a deviation which increases with an increase in the sweep rate. Also, the maximum amplitude during passage through resonance depends on the sweep rate. Ibrahim and Hijawi (1998) investigated the deterministic and stochastic response of a cantilever beam with a tip mass in the neighborhood of a combination parametric resonance. The results show that, for low excitation levels, the response is almost stationary and that its statistical parameters like the mean square, etc. possess unique values.

Nonresonant Modal Interactions

Haddow and Hasan (1988) observed an indirect excitation of a low-frequency mode when a cantilever beam was parametrically excited near twice its fourth natural frequency, which they described as an "extremely low subharmonic response." Burton and Kolowith (1988) conducted an experiment similar to that of Haddow and Hasan (1988). For certain excitation frequencies, they observed chaotic motions where the first seven in-plane bending modes as well as the first torsional mode were present in the response. Cusumano and Moon (1990,1995a,b) conducted an experiment with an externally excited cantilever beam, and they observed a cascading of energy into the low-frequency modes in the response associated with chaotic nonplanar motions.

In experiments with cantilever metallic beams, Anderson, Balachandran, and Nayfeh (1992,1994) and S. Nayfeh and Nayfeh (1994) displayed the transfer of energy between widely spaced modes. Anderson, Balachandran, and Nayfeh (1992,1994) conducted an experiment with a beam that was parametrically excited near twice its third-mode frequency. For certain excitation amplitudes and frequencies, they observed a large-amplitude planar motion consisting of the fourth, third, and first modes accompanied by amplitude and phase modulations of the third-mode response. S. Nayfeh and Nayfeh (1994) conducted an experiment with a circular metallic rod that was transversely excited near the natural frequency of its fifth mode. Because of axial symmetry, one-to-one internal resonances occur at each natural frequency of the beam, and the mode in the plane of excitation interacts with the out-of-plane mode of equal frequency, resulting in nonplanar whirling motions. For a certain range of parameters, they observed large-amplitude first-mode responses. As in the experiment of Anderson, Balachandran, and Nayfeh (1992,1994), the appearance of the first-mode response was accompanied by a modulation of the amplitude and phase of the fifth-mode response, with the modulation frequency being approximately equal to the natural frequency of the first mode. Smith, Balachandran, and Nayfeh (1992) examined the on-orbit data from the Hubble space telescope for indications of modal interactions. From the time history data, an energy transfer from high-frequency modes to a low-frequency mode is apparent. Also, the response spectra show sidebands which are indicative of modulated motions.

Popovic et al. (1995) demonstrated the transfer of energy between widely spaced modes in a three-beam frame structure with two corner masses, while Oh and Nayfeh (1998) experimentally documented such an energy transfer between a torsional mode and a bending mode of a stiff composite cantilever plate. Tabaddor and Nayfeh (1997) observed the same phenomenon with a cantilever steel beam externally excited around its fourth natural frequency. Exciting a horizontal metallic cantilever beam transversely near its first torsional frequency, Arafat and Nayfeh (1999) noted the activation of the first in-plane bending mode by a similar mechanism.

From the above experiments, conducted on various structures like a stiff composite plate, a portal frame (with dominant quadratic nonlinearities), and flexible cantilever beams (with dominant cubic nonlinearities), it is evident that energy transfer from a low-amplitude, high-frequency excitation to a high-amplitude, low-frequency response can occur in a structure irrespective of its stiffness, configuration, and inherent nonlinearities, as long as there exist modes whose natural frequencies are much lower than the natural frequencies of the modes being directly driven. In many engineering applications, high-frequency excitations can be caused by rotating machinery, and in space structures, for instance, there exists a large number of modes with very low natural frequencies.

S. Nayfeh and Nayfeh (1993) presented a paradigm for the transfer of energy among widely spaced modes in structures under external excitation. They studied a representative system made up of two nonlinearly coupled oscillators with cubic nonlinearities and widely spaced frequencies. S. Nayfeh (1993) studied a system comprising two nonlinearly coupled oscillators with quadratic nonlinearities and widely spaced frequencies. Results qualitatively similar to the experimental observations were obtained in both cases. S. Nayfeh (1993) also investigated the possibility of energy transfer from low- to high-frequency modes. Using a model similar to the one used to study the energy transfer from high- to low-frequency modes, S. Nayfeh (1993) found out that excitation of the low-frequency mode will not cause an energy transfer to the high-frequency mode. To date many experiments have been performed on the forced oscillations of the first few modes of continuous systems by many researchers. But, to the best of our knowledge, no report of the transfer of energy from low- to high-frequency modes exists in the literature.

Nayfeh and Chin (1995) extended the analysis of S. Nayfeh and Nayfeh (1993) to parametrically excited systems. They showed that exciting the high-frequency mode with a principal parametric resonance may result in a large-amplitude, low-frequency response accompanied by a slow modulation of the amplitude and phase of the high-frequency response, again similar to the experimental observations. Anderson, Nayfeh, and Balachandran (1996a) developed a theoretical model, based on the method of averaging, to study the energy transfer between widely spaced modes of a cantilever beam under parametric excitation. They obtained results that are in qualitative agreement with the experimental observations. Feng (1995) studied energy transfer from a high-frequency to a low-frequency mode in a mechanical system modeled as a linearly dissipative Hamiltonian system. Haller (1999) extended the work of Feng (1995) by including viscous damping and additional nonlinearity in the system and proved the existence of a Shilnikov-type orbit homoclinic to a saddle-focus.

Finally, it is noteworthy to cite a few books providing a comprehensive study of the various modal interactions and relevant review papers. Kármán (1940) gave a review of nonlinear problems in engineering along with their solutions. Bolotin (1964), Evan-Iwanowski (1976), Nayfeh (1973b,1981), Nayfeh and Mook (1979), and Nayfeh (2000) provided examples and detailed analyses of the various resonances and modal interactions in nonlinear systems. Rosenberg (1961) presented a short and instructive survey on nonlinear oscillations. Evan-Iwanowski (1965) gave a review of articles concerned with the parametric resonance in structures like beams, columns, arches, rings, plates, shells, etc. Wagner and Ramamurti (1977) presented a review of work done in the area of deterministic and random vibrations of beams using both linear and nonlinear theories. Sathyamoorthy (1982a) gave a survey of the literature on the nonlinear analysis of beams, considering both geometric and material nonlinearities. Nayfeh (1986) gave an overview of the perturbation methods used to obtain analytical solutions of nonlinear dynamical systems. Nayfeh and Balachandran (1989) presented a review of theoretical and experimental studies on modal interactions in dynamical and structural systems. Nayfeh and Mook (1995) presented a perspective of the mechanisms by which energy is transferred from high- to low-frequency modes. Nayfeh and Arafat (2001) gave an overview of nonlinear system dynamics.

1.3.4 System Identification

In recent years, identification of structural system models through the use of experimental data has received considerable attention owing to the increased importance given to the accurate prediction of the response of structures to various loading environments. The assumption that the effect of nonlinearity is negligible when the level of excitation is low is not always true. Also, the presence of nonlinearities in structures leads to many interesting physical phenomenon that cannot be explained by linear models (Nayfeh and Mook, 1979; Nayfeh, 2000). Therefore, emphasis is now on developing nonlinear system identification techniques that can predict those physical phenomena. Zavodney (1987) stressed the need to be aware and knowledgeable about nonlinear phenomena while estimating system parameters experimentally.

Nonlinear system identification techniques can be broadly classified into parametric and nonparametric methods. Parametric methods seek to determine the values of the parameters in an assumed model of the system to be identified, whereas nonparametric methods seek to determine the functional representation of the system to be identified. Most of the parametric and nonparametric identification methods employ, in one way or other, the least-squares approach in which the square of the error between the measured response and that of the identified model is minimized, thus providing the best estimate. Nonparametric identification methods are appropriate for systems whose model structures are unknown. The most commonly used nonparametric methods employ the Volterra series (Lee, 1997) and the restoring force-surface or force-state mapping method (Masri and Caughey, 1979; Masri, Sassi, and Caughey, 1982; Crawley and Aubert, 1986; Worden and Tomlinson, 2001). The Volterra series approach is computationally expensive, requires large storage space, has serious convergence problems, and cannot describe systems with multi-valued responses. In the restoring force-surface approach, simultaneous and accurate measurements of the input force and acceleration response are required. The corresponding displacement and velocity values are either obtained by direct measurements or through integration of the acceleration. Also, it is assumed that the restoring force is a function of the displacement and velocity only, which need not always be the case (Krauss and Nayfeh, 1999b).

Most of the parametric identification methods are time-domain based (Yasuda and Kamiya, 1999; Kapania and Park, 1997; Mohammad, Worden, and Tomlinson, 1992). The time-domain techniques have the advantages of requiring less time and effort for data acquisition than the sine-dwell method used for frequency-domain techniques and can be used for the identification of strongly nonlinear systems. Potential drawbacks of these approaches include problems of differentiating noisy signals and being unable to accurately estimate the coefficients of terms which are small. Frequency-domain techniques include approaches based on the backbone (or skeleton) curve and limit envelope (Tondl, 1975; Benhafsi, Penny, and Friswell, 1995; Fahey and Nayfeh, 1998), curve-fitting experimental frequency- and force-response data points (Krauss and Nayfeh, 1999a,b), the harmonic balance method (Yasuda, Kamiya, and Komakine, 1997), and methods exploiting nonlinear resonances (Nayfeh, 1985; Fahey and Nayfeh, 1998). Frequency-domain techniques avoid the problems associated with differentiation and observability of small terms, but require considerably more theoretical effort and are generally applicable only to weakly nonlinear systems.

1.3.5 Solution Methodologies

The equations of motion and the boundary conditions governing the nonlinear vibrations of a beam can be derived either by the Newtonian approach or by a variational approach. Hamilton's principle is the most widely used variational method. It is noteworthy to mention here that Hamilton's principle is a special case of Hamilton's law of varying action and can be used to obtain only the equations of motion and boundary conditions for dynamical problems. Hamilton's law, on the other hand, can be used to obtain approximate analytical solutions in addition to the equations of motion and boundary conditions (Bailey, 1975a,b). Having derived the governing equations and boundary conditions, the next step is to solve them. The principle of superposition, so commonly used in linear systems, is not applicable to nonlinear systems; thus making the determination of the response of nonlinear systems more difficult and equally challenging.

The nonlinear terms in the beam equations, for moderate rotations with small strains, are small compared to the linear terms. So, we restrict ourselves to the solution of weakly nonlinear continuous (distributed-parameter) systems. The equations of motion and/or boundary conditions, being nonlinear in nature, do not lend themselves to closed-form solutions. One, therefore, resorts to numerical techniques like the finite-element method or to approximate analytical techniques like discretization and/or perturbation methods. Here, we are primarily interested in the discussion of approximate analytical methods, and the method of multiple scales and the Galerkin discretization in particular. A review of the application of finite-element methods to the solution of nonlinear beam problems is given by Sathyamoorthy (1982b).

In the analysis of a weakly nonlinear continuous system, which has an infinite number of degrees of freedom, a modal discretization is often employed to obtain a reduced-order model of the system. By definition, a reduced-order model is a simplified mathematical model that encapsulates most, if not all, of the fundamental dynamics of a more complex system. In the discretization methods, one essentially postulates the system response in the form

$$v(s,t) = \sum_{n=1}^{N} q_n(t)\phi_n(s)$$

where N is a finite positive integer. Then, one assumes the spatial functions $\phi_n(s)$, space discretization, or the temporal functions $q_n(t)$, time discretization. With time discretization, the $q_n(t)$ are usually taken to be harmonic and the method of harmonic balance is used to obtain a set of nonlinear boundary-value problems for the $\phi_n(s)$. With space discretization, the $\phi_n(s)$ are assumed a priori. The $\phi_n(s)$ are usually taken to be the linear mode shapes. The method of weighted residuals or variational principles (like the Rayleigh-Ritz method) can then be used to obtain a reduced-order model comprising a set of ordinary-differential equations governing the modal coordinates $q_n(t)$, $n = 1, 2, \ldots, N$. The most popular implementation of weighted residuals is the Galerkin method, in which the trial functions $\phi_n(s)$ are also used as the weighing functions.

Perturbation techniques like the method of multiple scales are used to study the local dynamics of weakly nonlinear systems about an equilibrium state. But, unlike the discretization methods, they cannot be used to study the global dynamics of nonlinear systems. To obtain an approximate analytical solution of a weakly nonlinear continuous system, one can either directly apply a perturbation method to the governing partial-differential equations of motion and boundary conditions, or first discretize the partial-differential system to obtain a reduced-order model and then apply a perturbation method to the nonlinear ordinary-differential equations of the reduced-order model. The former procedure is usually referred to as the *direct* approach.

Application of the method of multiple scales, or any other perturbation method, to the reduced-order model, obtained by the Galerkin or other discretization procedures, of a weakly nonlinear continuous system with quadratic nonlinearities can lead to both quantitative and qualitative erroneous results (Pakdemirli, S. Nayfeh, and Nayfeh, 1995; Nayfeh and Lacarbonara, 1997; Alhazza and Nayfeh, 2001; Emam and Nayfeh, 2002; Nayfeh and Arafat, 2002). The direct approach is completely devoid of this problem. Also, such a problem does not exist for systems with just cubic nonlinearities. Lacarbonara (1999) showed that the quadratic nonlinearities produce a second-order contribution from all of the modes towards the system response in the case of a primary resonance. Hence, reduced-order discretization models may be inadequate to describe the dynamics of the original continuous system in the presence of quadratic nonlinearities. Nayfeh (1998) proposed a method for constructing reduced-order models of continuous systems with weak quadratic and cubic nonlinearities that overcomes this shortcoming of the discretization procedures.

Application of the method of multiple scales to dynamical systems expressed in second-order form can lead to modulation equations that cannot be derived from a Lagrangian in the absence of dissipation and external excitation, which is contrary to the conservative character of these dynamical systems. More specifically, this problem is encountered while determining approximate solutions of nonlinear systems possessing internal resonances to orders higher than the order at which the influence of the internal resonance first appears (Rega et al., 1999). Interestingly, transforming the second-order governing equations into a system of first-order equations and then treating them with the method of multiple scales yields modulation equations derivable from a Lagrangian (Nayfeh, 2000; Nayfeh and Chin, 1999).

1.4 Overview

In this dissertation, we study the nonlinear vibrations of metallic cantilever beams and plates subjected to transverse (or external or additive) harmonic excitation. The emphasis, however, is on the energy transfer between widely spaced modes via modulation and on parameter estimation. Both experimental and theoretical results are presented.

The equations of motion and boundary conditions governing the nonplanar, nonlinear vibrations of isotropic metallic beams are derived, using the extended Hamilton principle, in Chapter 2. Assumptions used in the derivation are also elaborated. The derived equations are used in the theoretical analysis done in subsequent chapters.

An experimental parametric identification technique to estimate the linear and nonlinear damping coefficients and effective nonlinearity of a metallic cantilever beam is presented in Chapter 3. This method is applicable to any single-degree-of-freedom nonlinear system with weak cubic geometric and inertia nonlinearities (Malatkar and Nayfeh, 2003a).

In Chapter 4, we propose two methods, based on the elimination theory of polynomials, which can be used to determine both the critical forcing amplitude as well as the jump frequencies in the case of single-degree-of-freedom nonlinear systems. The proposed methods have the potential of being applicable to multiple-degrees-of-freedom nonlinear systems (Malatkar and Nayfeh, 2002).

In Chapter 5, we investigate the transfer of energy between widely spaced modes via modulation in a flexible steel cantilever beam, of rectangular cross-section, under transverse excitation. This study is restricted to motion in a plane. We find experimentally that the energy transfer is very much dependent upon the closeness of the modulation frequency to the natural frequency of the first mode. A reduced-order analytical model is also developed to study the transfer of energy between widely spaced modes (Malatkar and Nayfeh, 2003b). In addition, we extend the planar reduced-order model to include out-of-plane modes and study the nonplanar energy transfer between widely spaced modes, in the presence of one-to-one internal resonance, in a circular steel rod under transverse excitation.

An experimental study of the response of a rectangular, aluminum cantilever plate to transverse harmonic excitations is presented in Chapter 6. It is shown that a simple cantilever plate can display a multitude of nonlinear dynamic phenomena (Malatkar and Nayfeh, 2003c). Also, we find again that the energy transfer between widely spaced modes via modulation is dependent upon the closeness of the modulation frequency to the first-mode natural frequency.

Chapter 2

Problem Formulation

In this chapter, we derive the equations of motion and boundary conditions governing the nonplanar, nonlinear vibrations of isotropic, inextensible, Euler-Bernoulli beams. Pai (1990) used a Newtonian approach to derive the nonlinear equations of motion describing the flexural-flexural-torsional vibrations of metallic and composite beams. Here we follow a variational approach, based on the extended Hamilton principle. In particular, we adopt the approach used by Crespo da Silva and Glynn (1978a,b) and Crespo da Silva (1988a). The simplifying assumptions and their validity are described as and when they are made during the derivation of the equations.

2.1 Beam Kinematics

A large deformation of a structure does not necessarily mean the presence of large strains. Under large rigid-body rotations, structures like cantilever beams undergo large deformations but small strains. Even when the rigid-body rotations are small, deformations will still be large for long structures. With respect to a coordinate system co-rotated with the rigid-body movement, the relative displacements are small and the problem becomes linearly elastic. But the large deformations give rise to geometric nonlinearities due to nonlinear curvature and/or midplane stretching, leading to nonlinear strain-displacement relations. Structures that undergo large deformations but small strains are labeled as nonlinear elastic structures (Nayfeh and Pai, 2003). We assume our beam to be a nonlinear elastic structure.

We use the Euler-Bernoulli beam theory to model the beam, and accordingly neglect the effects of warping and shear deformation. To simplify the expressions, we also neglect the usually small Poisson effect. For slender beams, these simplifications are valid. In the absence of warpings, a differential beam element can be considered as a rigid body, whose motion is then completely described by three translational and three rotational displacements. Also, knowing the deformation of the neutral axis (reference line) in space, one can determine the deformation of any other point on the beam.

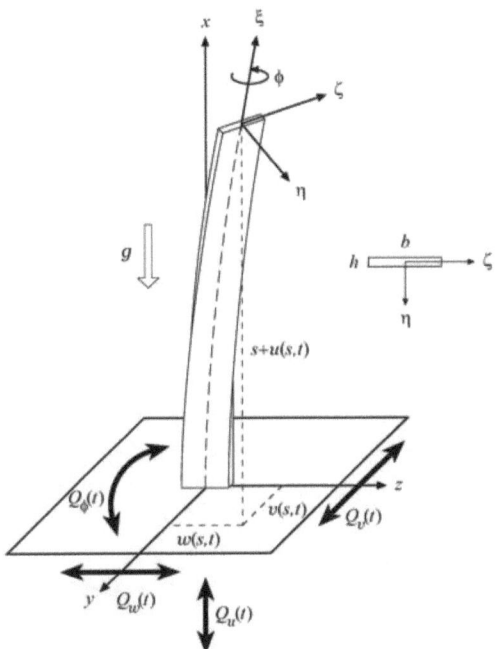

Figure 2.1: A schematic of a vertically mounted metallic cantilever beam undergoing flexural-flexural-torsional motions.

We consider a uniform and straight metallic cantilever beam of length l and mass per unit length m. A schematic of the beam is shown in Fig. 2.1, where (x, y, z) denote the inertial coordinate system with orthogonal unit vectors $(\mathbf{e}_x, \mathbf{e}_y, \mathbf{e}_z)$, while (ξ, η, ζ) denote a local curvilinear coordinate system at arclength s, in the deformed position, with orthogonal unit vectors $(\mathbf{e}_\xi, \mathbf{e}_\eta, \mathbf{e}_\zeta)$. As the beam has uniform cross sections and material properties, its mass and area centroids are identical and the principal axes of the beam's cross section at any s coincide with the (ξ, η, ζ) system. Moreover, the x and ξ axes represent the neutral axis of the beam before and after the deformation, respectively.

2.1.1 Euler-Angle Rotations

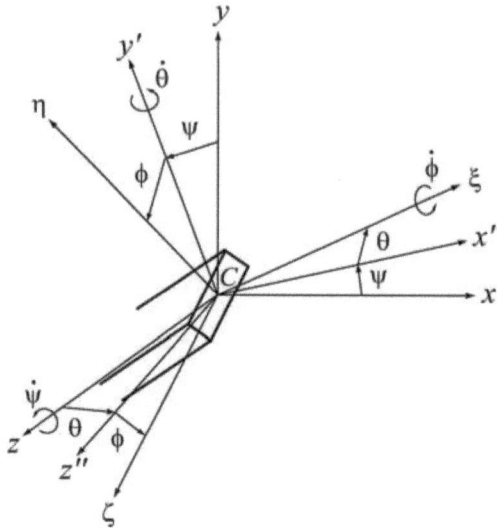

Figure 2.2: 3-2-1 Euler-angle rotations.

In general, each cross section of the beam experiences an elastic displacement of its centroid C and a rotation. The displacement components of the centroid C, with respect to the x, y, and z axes at the arclength s and time t are denoted by $u(s,t)$, $v(s,t)$, and $w(s,t)$, respectively, as shown in Fig. 2.1. To describe the rotation of the beam's cross section at C, from the undeformed to the deformed position, we use three successive Euler-angle (counterclockwise) rotations. Specifically, we use a 3-2-1 body rotation with the angles of rotation denoted, in the order of rotation, by $\psi(s,t)$, $\theta(s,t)$, and $\phi(s,t)$, as shown in Fig. 2.2. The first rotation ψ about $\mathbf{e}_z$ takes $(\mathbf{e}_x, \mathbf{e}_y, \mathbf{e}_z)$ to $(\mathbf{e}_{x'}, \mathbf{e}_{y'}, \mathbf{e}_{z'} = \mathbf{e}_z)$. The second rotation θ about $\mathbf{e}_{y'}$ takes $(\mathbf{e}_{x'}, \mathbf{e}_{y'}, \mathbf{e}_{z'})$ to $(\mathbf{e}_{x''}, \mathbf{e}_{y''} = \mathbf{e}_{y'}, \mathbf{e}_{z''})$, and the final rotation ϕ about $\mathbf{e}_{x''}$ takes

($\mathbf{e}_{x''}, \mathbf{e}_{y''}, \mathbf{e}_{z''}$) to the final orientation ($\mathbf{e}_\xi = \mathbf{e}_{x''}, \mathbf{e}_\eta, \mathbf{e}_\zeta$). The four unit-vector triads are related to each other in the following manner:

$$\begin{Bmatrix} \mathbf{e}_\xi \\ \mathbf{e}_\eta \\ \mathbf{e}_\zeta \end{Bmatrix} = [T_\phi] \begin{Bmatrix} \mathbf{e}_{x''} \\ \mathbf{e}_{y''} \\ \mathbf{e}_{z''} \end{Bmatrix} = [T_\phi][T_\theta] \begin{Bmatrix} \mathbf{e}_{x'} \\ \mathbf{e}_{y'} \\ \mathbf{e}_{z'} \end{Bmatrix} = \underbrace{[T_\phi][T_\theta][T_\psi]}_{=[T]} \begin{Bmatrix} \mathbf{e}_x \\ \mathbf{e}_y \\ \mathbf{e}_z \end{Bmatrix} \quad (2.1)$$

where

$$[T_\psi] = \begin{bmatrix} \cos\psi & \sin\psi & 0 \\ -\sin\psi & \cos\psi & 0 \\ 0 & 0 & 1 \end{bmatrix}, \quad [T_\theta] = \begin{bmatrix} \cos\theta & 0 & -\sin\theta \\ 0 & 1 & 0 \\ \sin\theta & 0 & \cos\theta \end{bmatrix}, \quad [T_\phi] = \begin{bmatrix} 1 & 0 & 0 \\ 0 & \cos\phi & \sin\phi \\ 0 & -\sin\phi & \cos\phi \end{bmatrix},$$

$$[T] = \begin{bmatrix} \cos\theta\cos\psi & \cos\theta\sin\psi & -\sin\theta \\ -\cos\phi\sin\psi + \sin\phi\sin\theta\cos\psi & \cos\phi\cos\psi + \sin\phi\sin\theta\sin\psi & \sin\phi\cos\theta \\ \sin\phi\sin\psi + \cos\phi\sin\theta\cos\psi & -\sin\phi\cos\psi + \cos\phi\sin\theta\sin\psi & \cos\phi\cos\theta \end{bmatrix} \quad (2.2)$$

The transformation matrices $[T_\psi]$, $[T_\theta]$, $[T_\phi]$, and $[T]$ are (proper) orthogonal or unitary matrices, and hence possess the property $[Q]^{-1} = [Q]^T$.

From Fig. 2.2, the absolute angular velocity $\boldsymbol{\omega}(s,t)$ of the principal axis system (ξ, η, ζ) can be obtained using Eqs. (2.1) and (2.2) as follows:

$$\begin{aligned} \boldsymbol{\omega}(s,t) &= \dot{\psi}\,\mathbf{e}_z + \dot{\theta}\,\mathbf{e}_{y'} + \dot{\phi}\,\mathbf{e}_\xi \\ &= (\dot{\phi} - \dot{\psi}\sin\theta)\,\mathbf{e}_\xi + (\dot{\psi}\cos\theta\sin\phi + \dot{\theta}\cos\phi)\,\mathbf{e}_\eta + (\dot{\psi}\cos\theta\cos\phi - \dot{\theta}\sin\phi)\,\mathbf{e}_\zeta \\ &\equiv \omega_\xi\,\mathbf{e}_\xi + \omega_\eta\,\mathbf{e}_\eta + \omega_\zeta\,\mathbf{e}_\zeta \end{aligned} \quad (2.3)$$

where the overdot stands for $\partial/\partial t$. According to the Kirchhoff's kinetic analogue (Love, 1944), the equations of a thin rod subjected only to end forces has the same form as those of a rigid body oscillating about a fixed point. Using the Kirchhoff's kinetic analogue, one can easily obtain expressions for the components of the curvature vector $\boldsymbol{\rho}(s,t)$ by simply replacing the time derivatives with the spatial derivatives in the angular velocity expression. Thus, from Eq. (2.3), we have

$$\begin{aligned} \boldsymbol{\rho}(s,t) &= (\phi' - \psi'\sin\theta)\,\mathbf{e}_\xi + (\psi'\cos\theta\sin\phi + \theta'\cos\phi)\,\mathbf{e}_\eta + (\psi'\cos\theta\cos\phi - \theta'\sin\phi)\,\mathbf{e}_\zeta \\ &\equiv \rho_\xi\,\mathbf{e}_\xi + \rho_\eta\,\mathbf{e}_\eta + \rho_\zeta\,\mathbf{e}_\zeta \end{aligned} \quad (2.4)$$

where the prime stands for $\partial/\partial s$. The curvature components can also be obtained using the definition of curvatures (Wempner, 1981; Nayfeh and Pai, 2003)

$$\rho_\xi \equiv \frac{\partial \mathbf{e}_\eta}{\partial s} \cdot \mathbf{e}_\zeta, \quad \rho_\eta \equiv -\frac{\partial \mathbf{e}_\xi}{\partial s} \cdot \mathbf{e}_\zeta, \quad \rho_\zeta \equiv \frac{\partial \mathbf{e}_\xi}{\partial s} \cdot \mathbf{e}_\eta \qquad (2.5)$$

where a dot denotes the inner product of two vectors. Using Eq. (2.1) in (2.5), we obtain Eq. (2.4).

2.1.2 Inextensional Beam

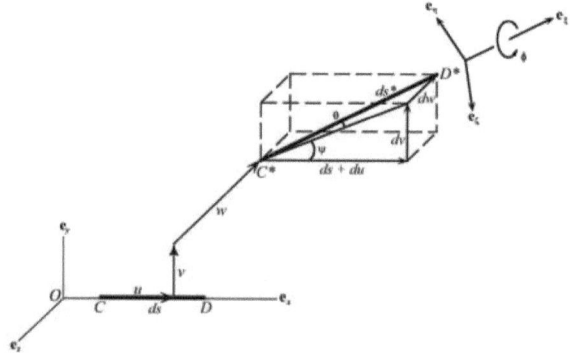

Figure 2.3: Deformation of a beam element along the neutral axis.

We now consider the deformation of an element CD of the beam's neutral axis, which is of length ds and located at a distance s from the origin O of the (x, y, z) system as shown in Fig. 2.3. Upon deformation, let CD move to C^*D^*. We denote the displacement components of C and D by (u, v, w) and $(u + du, v + dv, w + dw)$, respectively. From Fig. 2.3, the strain e at point C can be calculated as

$$e = \frac{ds^* - ds}{ds} = \sqrt{(1 + u')^2 + v'^2 + w'^2} - 1 \qquad (2.6)$$

We assume the beam's neutral axis to be inextensional; that is, $e = 0$. The inextensionality constraint equation thus is

$$(1 + u')^2 + v'^2 + w'^2 = 1 \qquad (2.7)$$

It is a well-known fact that, in the absence of large axial forces, fixed-free and fixed-sliding elements are approximated as inextensional members. The conditions under which this assumption is valid were discussed in the previous chapter.

Owing to the no-transverse-shear assumption, the rotation of the cross sections is due to bending alone. Therefore, from Fig. 2.3, we have

$$\tan\psi = \frac{v'}{1+u'}, \quad \tan\theta = \frac{-w'}{\sqrt{(1+u')^2 + v'^2}} \tag{2.8}$$

There is a minus sign in front of w' in the above equation because, for a counterclockwise rotation θ, $w' < 0$ as the w-displacement in such a case is along the negative z-axis.

2.1.3 Strain-Curvature Relations

It is known that rigid-body translations and rotations do not produce any strains; strains are only due to relative displacements. Next, we derive the expressions of the strain components by determining the deformation undergone by an infinitesimally small line segment.

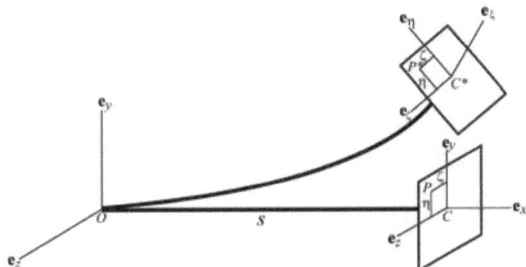

Figure 2.4: Initial and deformed positions of an arbitrary point P.

Consider the cross section of the beam at point C on the neutral axis, which is located at a distance s from the origin O of the inertial coordinate system (x, y, z), as shown in Fig. 2.4. Let P be a point on the cross section located at (η, ζ) relative to C. Upon deformation, let P move to P^*, with the displacement components of C being (u, v, w) in the (x, y, z) system. The coordinates of P^* relative to C^* are still (η, ζ) because of the assumption that the shape of the cross section remains intact after deformation. From Fig. 2.4, the position vectors of P and P^* can be written as

$$\boldsymbol{r}_P = \boldsymbol{OC} + \boldsymbol{CP} = s\,\boldsymbol{e}_x + \eta\,\boldsymbol{e}_y + \zeta\,\boldsymbol{e}_z \tag{2.9}$$

$$\boldsymbol{r}_{P^*} = \boldsymbol{OC^*} + \boldsymbol{C^*P^*} = (s+u)\,\boldsymbol{e}_x + v\,\boldsymbol{e}_y + w\,\boldsymbol{e}_z + \eta\,\boldsymbol{e}_\eta + \zeta\,\boldsymbol{e}_\zeta \tag{2.10}$$

Therefore,

$$dr_p = ds\,\mathbf{e}_x + d\eta\,\mathbf{e}_y + d\zeta\,\mathbf{e}_z \tag{2.11}$$

$$dr_{p^*} = (1+u')ds\,\mathbf{e}_x + v'ds\,\mathbf{e}_y + w'ds\,\mathbf{e}_z + d\eta\,\mathbf{e}_\eta + \eta\,d\mathbf{e}_\eta + d\zeta\,\mathbf{e}_\zeta + \zeta\,d\mathbf{e}_\zeta \tag{2.12}$$

From Fig. 2.3, using the inextensionality assumption, we have

$$\boldsymbol{C}^*\boldsymbol{D}^* = (1+u')ds\,\mathbf{e}_x + v'ds\,\mathbf{e}_y + w'ds\,\mathbf{e}_z = ds^*\,\mathbf{e}_\xi = ds\,\mathbf{e}_\xi \tag{2.13}$$

We know, for fixed s,

$$\frac{d\mathbf{e}_\alpha}{dt} = \boldsymbol{\omega} \times \mathbf{e}_\alpha \quad (\alpha = \xi, \eta, \zeta)$$

where $\times$ denotes the cross product of two vectors. Using Kirchhoff's kinetic analogue, for a given deformation (or fixed time t), we thus have

$$\frac{d\mathbf{e}_\alpha}{ds} = \boldsymbol{\rho} \times \mathbf{e}_\alpha \quad (\alpha = \xi, \eta, \zeta)$$

or $d\mathbf{e}_\alpha = (\boldsymbol{\rho} \times \mathbf{e}_\alpha)ds$. Using this equation along with Eq. (2.13) in Eq. (2.12), we obtain

$$dr_{p^*} = (1 + \zeta\rho_\eta - \eta\rho_\zeta)ds\,\mathbf{e}_\xi + (d\eta - \zeta\rho_\xi ds)\,\mathbf{e}_\eta + (d\zeta + \eta\rho_\xi ds)\,\mathbf{e}_\zeta \tag{2.14}$$

Using the definition of the Green's strain tensor ε (Fung, 1965; Annigeri, Cassenti, and Dennis, 1985; Shames and Dym, 1985), we have

$$dr_{p^*}\cdot dr_{p^*} - dr_p\cdot dr_p = 2\{ds\ d\eta\ d\zeta\}\,[\varepsilon_{ij}]\,\{ds\ d\eta\ d\zeta\}^T \quad (i,j=1,2,3) \tag{2.15}$$

where the ε_{ij} are components of the Green's strain tensor expressed in the Lagrangian (or undeformed) coordinates. From Eqs. (2.11) and (2.14), we now have

$$dr_{p^*}\cdot dr_{p^*} - dr_p\cdot dr_p = 2(\zeta\rho_\eta - \eta\rho_\zeta)\,ds^2 - 2\zeta\rho_\xi\,ds\,d\eta + 2\eta\rho_\xi\,ds\,d\zeta + H.O.T. \tag{2.16}$$

where H.O.T. stands for higher-order terms, which can be neglected as they are relatively small. Comparing Eqs. (2.15) and (2.16), we obtain

$$\varepsilon_{11} = \zeta\rho_\eta - \eta\rho_\zeta,\quad \gamma_{12} = 2\varepsilon_{12} = -\zeta\rho_\xi,\quad \gamma_{13} = 2\varepsilon_{13} = \eta\rho_\xi,\quad \varepsilon_{22} = \varepsilon_{23} = \varepsilon_{33} = 0 \tag{2.17}$$

where γ_{12} and γ_{13} denote engineering shear strains. We note that only shear strains due to torsion are accounted for. The Euler-Bernoulli beam assumptions of no-transverse-shear and no strains in the plane of the cross section result in the corresponding strain components being equal to zero. For a slender beam undergoing moderate rotations, $\eta, \zeta, \rho_\eta, \rho_\zeta, \rho_\xi$ are small, and thus the strain components, given by Eq. (2.17), are also small compared to unity (Wempner, 1981).

2.2 Equations of Motion

The variables ψ and θ are dependent variables, as seen in Eq. (2.8). Therefore, the number of independent variables now is reduced to four; namely, u, v, w, and ϕ. For an inextensional beam, the number of independent variables can be reduced, using the inextensionality constraint, to three; namely, v, w, and ϕ. Assuming the rotatory inertia to be negligible compared to the translational inertia, we can also get rid of ϕ, leaving only v and w. In this section, we develop the two nonlinear partial-differential equations of motion describing the flexural-flexural motion of an inextensional beam. The extended Hamilton principle is used to derive these equations, and in the process the inextensionality constraint will be clubbed to the Lagrangian of the motion through a Lagrange multiplier.

2.2.1 Lagrangian of Motion

The Lagrangian of motion $\mathcal{L}$ is defined as

$$\mathcal{L} \equiv T - V = \int_0^l \ell \, ds \tag{2.18}$$

where T is the kinetic energy, V is the potential energy, l denotes the length of the beam, and ℓ is the specific (i.e., per unit length) Lagrangian.

The kinetic energy of the beam consists of two parts – translational and rotational. The translational kinetic energy is given by

$$T_1 = \frac{1}{2} m \int_0^l (\dot{u}^2 + \dot{v}^2 + \dot{w}^2) \, ds \tag{2.19}$$

and the rotational kinetic energy is given by

$$T_2 = \frac{1}{2} \int_0^l \{\omega_\xi \;\; \omega_\eta \;\; \omega_\zeta\} [J] \{\omega_\xi \;\; \omega_\eta \;\; \omega_\zeta\}^T \, ds \tag{2.20}$$

where $[J]$ is the distributed inertia matrix. Because the local coordinate axes coincide with the principal axes of the beam, we have

$$[J] = \begin{bmatrix} J_\xi & 0 & 0 \\ 0 & J_\eta & 0 \\ 0 & 0 & J_\zeta \end{bmatrix}$$

where J_ξ, J_η, and J_ζ are the principal mass moments of inertia per unit length of the beam, they are defined as

$$J_\xi = \iint_A \rho(\eta^2 + \zeta^2) \, d\eta \, d\zeta, \quad J_\eta = \iint_A \rho\zeta^2 \, d\eta \, d\zeta, \quad J_\zeta = \iint_A \rho\eta^2 \, d\eta \, d\zeta$$

Here, ρ denotes the mass density of the beam and A denotes the area of the cross section of the beam located at a distance s from the origin of the (x,y,z) system. As the beam is uniform, J_ξ, J_η, and J_ζ are constants. The total kinetic energy can now be written as

$$T = \frac{1}{2} \int_0^l \left[m(\dot{u}^2 + \dot{v}^2 + \dot{w}^2) + J_\xi \omega_\xi^2 + J_\eta \omega_\eta^2 + J_\zeta \omega_\zeta^2 \right] ds \qquad (2.21)$$

The potential energy V can be determined from the corresponding strain energy U, which for our beam is given by

$$U = \frac{1}{2} \int_0^l \left\{ \iint_A (\sigma_{11}\varepsilon_{11} + \sigma_{12}\gamma_{12} + \sigma_{13}\gamma_{13}) \, d\eta \, d\zeta \right\} ds \qquad (2.22)$$

where the σ_{ij} denote components of the stress tensor. To arrive at Eq. (2.22), we used the fact that the beam is an elastic structure with a linear stress-strain relationship and that $\varepsilon_{22} = \varepsilon_{33} = \gamma_{23} = 0$. Using Hooke's law and neglecting Poisson's effect (to simplify the expressions of the stress components), we can write $\sigma_{11} \approx E\varepsilon_{11}$, $\sigma_{12} \approx G\gamma_{12}$, $\sigma_{13} \approx G\gamma_{13}$, where E and G are the Young's and shear moduli, respectively, of the beam. Using these relations and Eq. (2.17) in Eq. (2.22), we obtain

$$V = U = \frac{1}{2} \int_0^l \left\{ \iint_A \left[E(\zeta\rho_\eta - \eta\rho_\zeta)^2 + G\zeta^2\rho_\xi^2 + G\eta^2\rho_\xi^2 \right] d\eta \, d\zeta \right\} ds \qquad (2.23)$$

Using the fact that the curvature components are not functions of η or ζ and that the cross section is symmetric about the η and ζ axes, we reduce Eq. (2.23) to

$$V = \frac{1}{2} \int_0^l \left(D_\xi \rho_\xi^2 + D_\eta \rho_\eta^2 + D_\zeta \rho_\zeta^2 \right) ds \qquad (2.24)$$

where

$$D_\xi = G \iint_A (\eta^2 + \zeta^2) \, d\eta \, d\zeta, \quad D_\eta = E \iint_A \zeta^2 \, d\eta \, d\zeta, \quad D_\zeta = E \iint_A \eta^2 \, d\eta \, d\zeta$$

are the torsional and bending stiffnesses, respectively, of the beam.

To enforce the inextensionality constraint, we use the Lagrange multiplier $\lambda(s,t)$. Using Eqs. (2.21) and (2.24) and the inextensionality constraint given by Eq. (2.7), we write the overall specific Lagrangian ℓ as follows:

$$\ell = \frac{1}{2} m(\dot{u}^2 + \dot{v}^2 + \dot{w}^2) + \frac{1}{2}(J_\xi \omega_\xi^2 + J_\eta \omega_\eta^2 + J_\zeta \omega_\zeta^2) - \frac{1}{2}(D_\xi \rho_\xi^2 + D_\eta \rho_\eta^2 + D_\zeta \rho_\zeta^2)$$
$$+ \frac{1}{2}\lambda \left[1 - (1+u')^2 - v'^2 - w'^2 \right] \qquad (2.25)$$

The expressions for the components of $\boldsymbol{\omega}$ and $\boldsymbol{\rho}$, appearing in the above equation, are given in Eqs. (2.3) and (2.4).

2.2.2 Extended Hamilton Principle

Hamilton's principle (Meirovitch, 1967) states that, of all the varied paths satisfying the prescribed initial and final configurations, the actual (or true) path extremizes the functional $I = \int_{t_1}^{t_2} \mathcal{L}\, dt$, where t_1 and t_2 denote the initial and final time instants. By also including the work done by non-conservative forces within the integrand, we get the extended Hamilton principle. Using the variation of the functional I and the fact that the variation and integral operators commute, we can write for the actual path

$$\delta I = \int_{t_1}^{t_2} (\delta \mathcal{L} + \delta W_{nc})\, dt = 0 \qquad (2.26)$$

where W_{nc} denotes the work done by non-conservative forces, such as damping, external forces and moments. This condition of stationarity leads to all of the equations of motion and boundary conditions. Using the generalized forces along the x, y, z, and ϕ axes denoted by Q_u, Q_v, Q_w, and Q_ϕ, respectively, and the corresponding damping coefficients denoted by c_u, c_v, c_w, and c_ϕ, respectively, we write the expression for δW_{nc} as

$$\begin{aligned}\delta W_{nc} &= \int_0^l \left[(Q_u - c_u \dot{u})\delta u + (Q_v - c_v \dot{v})\delta v + (Q_w - c_w \dot{w})\delta w + (Q_\phi - c_\phi \dot{\phi})\delta\phi\right] ds & (2.27)\\ &\equiv \int_0^l \left[Q_u^* \delta u + Q_v^* \delta v + Q_w^* \delta w + Q_\phi^* \delta\phi\right] ds & (2.28)\end{aligned}$$

Substituting Eqs. (2.25) and (2.28) into Eq. (2.26), we obtain

$$\delta I = \int_{t_1}^{t_2} \int_0^l \left[\delta \ell + Q_u^* \delta u + Q_v^* \delta v + Q_w^* \delta w + Q_\phi^* \delta \phi\right] ds\, dt = 0 \qquad (2.29)$$

The specific Lagrangian ℓ is a function of x_i $(i = 1, 2, \ldots, 13)$ where $\boldsymbol{x} = \{\dot{u}, \dot{v}, \dot{w}, \psi, \theta, \phi, \dot{\psi}, \dot{\theta}, \dot{\phi}, \psi', \theta', \phi', \lambda\}^T$. Therefore,

$$\delta \ell = \sum_{i=1}^{13} \frac{\partial \ell}{\partial x_i} \delta x_i \qquad (2.30)$$

But there are only four independent variables, namely, u, v, w, and ϕ. Variations of the dependent variables ψ and θ can be obtained using Eq. (2.8), and are given by

$$\delta\psi = \frac{\partial\psi}{\partial u'}\delta u' + \frac{\partial\psi}{\partial v'}\delta v' = \frac{-v'\delta u' + (1+u')\delta v'}{(1+u')^2 + v'^2} \tag{2.31}$$

$$\delta\theta = \frac{\partial\theta}{\partial u'}\delta u' + \frac{\partial\theta}{\partial v'}\delta v' + \frac{\partial\theta}{\partial w'}\delta w' = \frac{w'[(1+u')\delta u' + v'\delta v']}{\sqrt{(1+u')^2 + v'^2}} - \sqrt{(1+u')^2 + v'^2}\,\delta w' \tag{2.32}$$

The variation and derivative operators commute. Thus, the variations $\delta\dot\alpha$, $\delta\alpha'$ ($\alpha = \psi, \theta$) can be written as $\frac{\partial}{\partial t}(\delta\alpha)$, $\frac{\partial}{\partial s}(\delta\alpha)$ ($\alpha = \psi, \theta$), respectively.

Substituting Eqs. (2.31) and (2.32) into Eq. (2.30), substituting the result in turn into Eq. (2.29), and after performing a few integrations by parts in Eq. (2.29), we obtain

$$\int_{t_1}^{t_2}\left\{\int_0^l(-m\ddot{u} + Q_u^* + G_u')\delta u\,ds + \int_0^l(-m\ddot{v} + Q_v^* + G_v')\delta v\,ds + \right.$$
$$\int_0^l(-m\ddot{w} + Q_w^* + G_w')\delta w\,ds + \int_0^l(Q_\phi^* - A_\phi)\delta\phi\,ds + $$
$$\left.\left[-G_u\delta u - G_v\delta v - G_w\delta w + H_u\delta u' + H_v\delta v' + H_w\delta w' + \frac{\partial\ell}{\partial\phi'}\delta\phi\right]\bigg|_{s=0}^l \right\}dt = 0 \tag{2.33}$$

where

$$G_u = A_\psi\frac{\partial\psi}{\partial u'} + A_\theta\frac{\partial\theta}{\partial u'} + \lambda(1+u')$$
$$G_v = A_\psi\frac{\partial\psi}{\partial v'} + A_\theta\frac{\partial\theta}{\partial v'} + \lambda v'$$
$$G_w = A_\theta\frac{\partial\theta}{\partial w'} + \lambda w'$$

and

$$A_\alpha = \frac{\partial^2\ell}{\partial t\,\partial\dot\alpha} + \frac{\partial^2\ell}{\partial s\,\partial\alpha'} - \frac{\partial\ell}{\partial\alpha} \quad (\alpha = \psi, \theta, \phi)$$
$$H_\alpha = \frac{\partial\ell}{\partial\psi'}\frac{\partial\psi}{\partial\alpha'} + \frac{\partial\ell}{\partial\theta'}\frac{\partial\theta}{\partial\alpha'} \quad (\alpha = u, v, w)$$

Equation (2.33) is valid for any arbitraty δu, δv, δw, and $\delta\phi$, implying that the individual integrands be equal to zero. Therefore,

$$m\ddot{u} - Q_u^* = G_u' \tag{2.34}$$
$$m\ddot{v} - Q_v^* = G_v' \tag{2.35}$$
$$m\ddot{w} - Q_w^* = G_w' \tag{2.36}$$
$$Q_\phi^* = A_\phi \tag{2.37}$$

and

$$\left[-G_u\delta u - G_v\delta v - G_w\delta w + H_u\delta u' + H_v\delta v' + H_w\delta w' + \frac{\partial \ell}{\partial \phi'}\delta\phi\right]\bigg|_{s=0}^{l} = 0 \quad (2.38)$$

Alternatively, the above equations of motion and boundary conditions could have been derived using Lagrange's equation for distributed systems (Meirovitch, 1997). Arafat (1999) used such an approach to obtain the equations of motion and boundary conditions describing the nonlinear vibrations of metallic and symmetrically laminated composite beams.

2.2.3 Order-Three Equations of Motion

Equations (2.34)-(2.38) are valid for arbitrarily large deformations as long as strains are small. But the two-point nonlinear boundary-value problem is not amenable for a closed-form solution because the equations are transcendental. One approach would be to resort to direct numerical procedures, but they suffer from instability and convergence problems. Another approach would be to expand the nonlinear transcendental terms into polynomials. Here, we expand those nonlinear terms into polynomials of order three. The third-order nonlinear equations of motion thus obtained would be appropriate for analyzing small but finite oscillations about the equilibrium (or undeformed) position.

We assume that v, w, and their derivatives are $O(\epsilon)$, where ϵ ($\ll 1$) is a bookkeeping parameter that is introduced to keep track of the different orders of approximation. We now expand all terms in Eqs. (2.34)-(2.38) in Taylor series and keep nonlinear terms up to $O(\epsilon^3)$.

We know the Taylor series expansion of $\tan^{-1} x$ (or $\arctan x$), up to order three, is given by

$$\tan^{-1} x = x - \frac{1}{3}x^3 + \cdots \quad (2.39)$$

Using Eqs. (2.7), (2.8), and (2.39), we obtain

$$u' = (1 - v'^2 - w'^2)^{1/2} - 1 = -\frac{1}{2}(v'^2 + w'^2) + \cdots \quad (2.40)$$

$$\psi = \tan^{-1}\frac{v'}{1+u'} = \tan^{-1}\left[v'(1 - v'^2 - w'^2)^{-1/2}\right] = v'\left(1 + \frac{1}{6}v'^2 + \frac{1}{2}w'^2\right) + \cdots \quad (2.41)$$

$$\theta = \tan^{-1}\frac{-w'}{[(1+u')^2 + v'^2]^{1/2}} = \tan^{-1}\left[-w'(1 - w'^2)^{-1/2}\right] = -w'\left(1 + \frac{1}{6}w'^2\right) + \cdots \quad (2.42)$$

In the nonlinear formulation, ϕ does not physically represent the real angle of twist with respect to the beam's axis. The third-order expansion of the twisting curvature ρ_ξ can be obtained using

Eqs. (2.41) and (2.42) in (2.4), and is given by $\rho_\xi = \phi' + v''w'$. Thus, a non-zero ϕ does not necessarily indicate the presence of torsion along the beam (Pai, 1990). For the 3-2-1 body rotation sequence, we define the twist angle γ as

$$\gamma \equiv \phi + \int_0^s v'' w' \, ds \tag{2.43}$$

Thus, $\rho_\xi = \gamma'$.

Next, we consider beams whose torsional rigidity is relatively high compared to the flexural rigidity. This is true for long beams with near-circular or near-square cross sections. In such a case, the torsional inertia cannot be excited by low-frequency excitations because the fundamental torsional frequency is much higher than the frequencies of the directly excited flexural modes. In addition, we assume that the distributed mass moments of inertia of the beam exert a negligible influence on its motion. In other words, the rotatory inertia is considered to be small compared to the translational inertia. This is a valid assumption for slender beams. Using Eqs. (2.40)-(2.42) and (2.43) in Eqs. (2.34)-(2.38), dropping terms containing J_ξ, J_η, and J_ζ, setting $Q_\phi^* = 0$, and retaining nonlinearities up to order three, we obtain

$$m\ddot{u} + c_u \dot{u} - Q_u = \Big\{ D_\xi \gamma'(w''v' - v''w') - (D_\eta - D_\zeta)[w'(v''\gamma)' + v'(w''\gamma)'] \\ + D_\zeta v''' v' + D_\eta w''' w' + \lambda(1 + u') \Big\}' \tag{2.44}$$

$$m\ddot{v} + c_v \dot{v} - Q_v = \Big\{ -D_\xi \gamma' w'' + (D_\eta - D_\zeta)\left[(w''\gamma)' - (v''\gamma^2)' + w''' \int_0^s v'w'' \, ds\right] \\ - D_\zeta [v''' + v'(v''^2 + w''^2)] + \lambda v' \Big\}' \tag{2.45}$$

$$m\ddot{w} + c_w \dot{w} - Q_w = \Big\{ D_\xi \gamma' v'' + (D_\eta - D_\zeta)\left[(v''\gamma)' + (w''\gamma^2)' - v''' \int_0^s w'v'' \, ds\right] \\ - D_\eta [w''' + w'(v''^2 + w''^2)] + \lambda w' \Big\}' \tag{2.46}$$

$$D_\xi \gamma'' = (D_\eta - D_\zeta)\left[\gamma(v''^2 - w''^2) - v''w''\right] \tag{2.47}$$

and the associated boundary conditions now become

$$\alpha(0,t) = 0 \quad (\alpha = u, v, w, \gamma, v', w') \tag{2.48}$$

$$\alpha(l,t) = 0 \quad \left(\alpha = H_v - H_u \frac{v'}{1+u'}, H_w - H_u \frac{w'}{1+u'}, \gamma'\right) \tag{2.49}$$

$$G_\alpha(l,t) = 0 \quad (\alpha = u, v, w) \tag{2.50}$$

From Eqs. (2.44) and (2.47), it is clear that u, λ, and γ are $O(\epsilon^2)$. Thus, for a weakly damped system like our beam, the damping terms $c_u \dot{u}$ and $c_\phi \dot{\phi}$ turn out to be very small, and hence they can be dropped from the equations of motion. Also, we see that if $D_\eta = D_\zeta$, then there will be no coupling between the flexural and torsional motions. In fact, in such a case, $\gamma = 0$.

Using the boundary conditions $u(0,t) = 0$, $G_u(L,t) = 0$, $\gamma(0,t) = 0$, and $\gamma'(L,t) = 0$ in Eqs. (2.40), (2.44), and (2.47), we obtain

$$u = -\frac{1}{2}\int_0^s (v'^2 + w'^2)\,ds \tag{2.51}$$

$$\lambda = -D_\zeta v''' v' - D_\eta w''' w' - \frac{1}{2}m\int_l^s \left[\int_0^s (v'^2 + w'^2)\,ds\right]^{\cdot\cdot}\,ds - \int_l^s Q_u\,ds \tag{2.52}$$

$$\gamma = -\frac{D_\eta - D_\zeta}{D_\xi}\int_0^s\int_l^s v'' w''\,ds\,ds \tag{2.53}$$

Equation (2.53) shows that the bending-induced twisting is a nonlinear phenomenon. The Lagrange multiplier $\lambda(s,t)$ is interpreted as an axial force, necessary to maintain the inextensionality constraint.

Substituting Eqs. (2.51)-(2.53) into Eqs. (2.45), (2.46), and (2.48)-(2.50) and keeping terms up to order three, we obtain

$$m\ddot{v} + c_v\dot{v} + D_\zeta v^{iv} = Q_v + \left\{(D_\eta - D_\zeta)\left[w''\int_l^s v'' w''\,ds - w'''\int_0^s v'' w'\,ds\right]\right.$$

$$\left. - \frac{(D_\eta - D_\zeta)^2}{D_\xi}\left(w''\int_0^s\int_l^s v'' w''\,ds\,ds\right)'\right\} - D_\zeta\left\{v'(v'v'' + w'w'')'\right\}'$$

$$-\frac{1}{2}m\left\{v'\int_l^s\left[\int_0^s (v'^2 + w'^2)\,ds\right]^{\cdot\cdot}\,ds\right\}' - \left(v'\int_l^s Q_u\,ds\right)' \tag{2.54}$$

$$m\ddot{w} + c_w\dot{w} + D_\eta w^{iv} = Q_w - \left\{(D_\eta - D_\zeta)\left[v''\int_l^s v'' w''\,ds - v'''\int_0^s w'' v'\,ds\right]\right.$$

$$\left. + \frac{(D_\eta - D_\zeta)^2}{D_\xi}\left(v''\int_0^s\int_l^s v'' w''\,ds\,ds\right)'\right\} - D_\eta\left\{w'(v'v'' + w'w'')'\right\}'$$

$$-\frac{1}{2}m\left\{w'\int_l^s\left[\int_0^s (v'^2 + w'^2)\,ds\right]^{\cdot\cdot}\,ds\right\}' - \left(w'\int_l^s Q_u\,ds\right)' \tag{2.55}$$

with the boundary conditions now being

$$v(0,t) = 0,\ w(0,t) = 0,\ v'(0,t) = 0,\ w'(0,t) = 0 \tag{2.56}$$

$$v''(l,t) = 0,\ w''(l,t) = 0,\ v'''(l,t) = 0,\ w'''(l,t) = 0 \tag{2.57}$$

In the above equations of motion, only cubic nonlinearities are present. The nonlinear term on the right-hand side of Eqs. (2.54) and (2.55), with the time derivatives, is the inertia nonlinearity arising from the kinetic energy of axial motion. The rest of the nonlinear terms are of the geometric nonlinearity type and originate from the potential energy stored in bending.

When the beam is subjected only to a transverse base excitation in the y-direction, with all other external forces except gravity being absent, we have $Q_v = Q_w = 0$, $Q_u = -mg$, and $v = \bar{v} + v_0 \cos(\Omega t)$, where g ($= 9.8\ m/s^2$) denotes the acceleration due to gravity, $\bar{v}$ is the displacement in the y-direction, with respect to the base, and v_0 and Ω are the amplitude and frequency of the base motion. Also, δv should be replaced by $\delta \bar{v}$ wherever it appears in the above equations, and in Eq. (2.27), $c_v \dot{v}$ should be replaced by $c_v \dot{\bar{v}}$. In Eq. (2.54), $m\ddot{v} = m\ddot{\bar{v}} - mv_0\Omega^2 \cos(\Omega t) = m\ddot{\bar{v}} - ma_b \cos(\Omega t)$, where a_b denotes the amplitude of the base acceleration. The equations of motion and boundary conditions now become

$$\begin{aligned}
m\ddot{v} + c_v \dot{v} + D_\zeta v^{iv} &= mg[v''(s-l) + v']' + \left\{ (D_\eta - D_\zeta) \left[w'' \int_l^s v''w''\,ds - w''' \int_0^s v''w'\,ds \right] \right. \\
&\quad \left. - \frac{(D_\eta - D_\zeta)^2}{D_\xi} \left(w'' \int_0^s \int_l^s v''w''\,ds\,ds \right)' \right\}' - D_\zeta \{ v'(v'v'' + w'w'')' \}' \\
&\quad - \frac{1}{2} m \left\{ v' \int_l^s \left[\int_0^s (v'^2 + w'^2)\,ds \right]^{\cdot\cdot} ds \right\}' + ma_b \cos(\Omega t) \quad (2.58)
\end{aligned}$$

$$\begin{aligned}
m\ddot{w} + c_w \dot{w} + D_\eta w^{iv} &= mg[w''(s-l) + w']' - \left\{ (D_\eta - D_\zeta) \left[v'' \int_l^s v''w''\,ds - v''' \int_0^s w''v'\,ds \right] \right. \\
&\quad \left. + \frac{(D_\eta - D_\zeta)^2}{D_\xi} \left(v'' \int_0^s \int_l^s v''w''\,ds\,ds \right)' \right\}' - D_\eta \{ w'(v'v'' + w'w'')' \}' \\
&\quad - \frac{1}{2} m \left\{ w' \int_l^s \left[\int_0^s (v'^2 + w'^2)\,ds \right]^{\cdot\cdot} ds \right\}' \quad (2.59)
\end{aligned}$$

$$v(0,t) = 0,\ w(0,t) = 0,\ v'(0,t) = 0,\ w'(0,t) = 0 \quad (2.60)$$

$$v''(l,t) = 0,\ w''(l,t) = 0,\ v'''(l,t) = 0,\ w'''(l,t) = 0 \quad (2.61)$$

where the bar over v has been dropped for ease of notation.

Chapter 3

Parametric System Identification

In this chapter, we propose a simple parametric identification technique for single-degree-of-freedom (SDOF) nonlinear systems with weak cubic nonlinearities. The proposed technique is related to the backbone curve method in the sense that it also uses the peak of the frequency-response curve of the nonlinear system to estimate the model parameters. But the proposed technique is much more simple and straightforward compared to the backbone curve method. The proposed identification procedure is outlined in the context of a single-mode response of an externally excited cantilever beam possessing cubic geometric and inertia nonlinearities and linear and quadratic damping.

3.1 Theoretical Modeling

3.1.1 Equation of Motion

Equations (2.58) and (2.59) governing the nonplanar dynamics of an isotropic, inextensional beam are simplified to the case of planar motion of a uniform metallic cantilever beam under external excitation. Following the approach of Anderson, Nayfeh, and Balachandran (1996b) and Tabaddor (2000), we also include quadratic damping (air drag) in the model, in addition to linear damping, to study its influence on the beam response. For the natural frequencies, we use the experimental values instead of the theoretical ones, and thus drop the gravity term from the equation of motion. The model equation used in this study is as follows:

$$m\ddot{v} + c_v \dot{v} + \widehat{c}\dot{v}|\dot{v}| + EIv^{iv} = ma_b \cos(\Omega t) - EI\left[v'(v'v'')'\right]' - \frac{1}{2}m\left\{v'\int_l^s \left[\frac{\partial^2}{\partial t^2}\int_0^s v'^2 ds\right]ds\right\}' \quad (3.1)$$

and the boundary conditions are

$$v(0,t) = 0, \ v'(0,t) = 0 \quad (3.2)$$

$$v''(l,t) = 0, \ v'''(l,t) = 0 \quad (3.3)$$

where m is the beam mass per unit length, l is the beam length, E is Young's modulus, I is the area moment of inertia, s is the arclength, t is time, $v(s,t)$ is the transverse displacement, a_b is the acceleration of the supported end of the beam, c_v is the coefficient of linear viscous damping per unit length, $\widehat{c}$ is the coefficient of quadratic damping per unit length, and Ω is the excitation frequency. And, the prime indicates differentiation with respect to the arclength s, whereas the overdot indicates differentiation with respect to time t.

3.1.2 Single-Mode Response

The steel beam used in the experiments constitutes a lightly damped, weakly nonlinear system, and none of its modes is involved in an internal resonance with other modes. We, therefore, assume that the response of the beam consists essentially of the undamped linear mode whose natural frequency is closest to the excitation frequency. We refer to this mode as the nth mode whose frequency ω_n is then very close to the excitation frequency Ω. Other modes, not being directly or indirectly excited, will decay to zero with time due to the presence of damping (Nayfeh and Mook, 1979).

Equations (3.1)-(3.3) are not readily amenable to a closed-form solution. We, therefore, resort to perturbation methods to obtain an approximate analytical solution. The method of multiple scales (Nayfeh, 1973b,1981) is used to derive a first-order uniform expansion for the beam response under primary resonance. Using a method of multiple scales' model for system identification would lead to biased parameter estimates at high levels of excitation (Doughty, Davies, and Bajaj, 2002). In the experiments, the excitation levels are kept low and so we need not be unduly concerned about any bias creeping into the estimates.

We scale the damping coefficients c_v and $\widehat{c}$ and the forcing coefficient a_b appearing in Eq. (3.1) in terms of a small dimensionless parameter ϵ ($\ll 1$) as follows:

$$\frac{c_v}{2m} = \zeta \omega_n = \epsilon^2 \mu \tag{3.4}$$

$$\frac{\hat{c}}{m} = \bar{c} = \epsilon c \tag{3.5}$$

$$a_b = \epsilon^3 \hat{f} \tag{3.6}$$

where ζ is the dimensionless linear viscous damping factor corresponding to the nth mode. Also, we let

$$v(s,t;\epsilon) = \epsilon\, v_1(s,T_0,T_2) + \epsilon^3 v_3(s,T_0,T_2) + \cdots \tag{3.7}$$

where the T_n ($=\epsilon^n t$) represent different time scales – T_0 being the fast-time scale and T_2 the slow-time scale. The derivatives with respect to t now take the form

$$\frac{d}{dt} = D_0 + \epsilon D_1 + \epsilon^2 D_2 + \cdots \tag{3.8}$$

$$\frac{d^2}{dt^2} = D_0^2 + 2\epsilon D_0 D_1 + \epsilon^2(D_1^2 + 2D_0 D_2) + \cdots \tag{3.9}$$

where $D_n = \partial/\partial T_n$. Substituting Eqs. (3.4)-(3.9) into Eqs. (3.1)-(3.3) and equating coefficients of like powers of ϵ, we obtain

Order ϵ:

$$D_0^2 v_1 + \frac{EI}{m} v_1^{iv} = 0 \tag{3.10}$$

$$v_1 = 0 \text{ and } v_1' = 0 \text{ at } s = 0 \tag{3.11}$$

$$v_1'' = 0 \text{ and } v_1''' = 0 \text{ at } s = l \tag{3.12}$$

Order ϵ^3:

$$D_0^2 v_3 + \frac{EI}{m} v_3^{iv} = -2D_0 D_2 v_1 - 2\mu D_0 v_1 - c D_0 v_1 \mid D_0 v_1 \mid -\frac{EI}{m}\left[v_1'\left(v_1' v_1''\right)'\right]'$$
$$-\frac{1}{2}\left[v_1' \int_l^s D_0^2 \left(\int_0^s v_1'^2 ds\right) ds\right]' + \hat{f}\cos(\Omega T_0) \tag{3.13}$$

$$v_3 = 0 \text{ and } v_3' = 0 \text{ at } s = 0 \tag{3.14}$$

$$v_3'' = 0 \text{ and } v_3''' = 0 \text{ at } s = l \tag{3.15}$$

Since we are seeking a single-mode response solution, the solution of the first-order problem associated with Eqs. (3.10)-(3.12) is taken as

$$v_1(s, T_0, T_2) = \left(A(T_2)e^{i\omega_n T_0} + \bar{A}(T_2)e^{-i\omega_n T_0}\right)\Phi_n(s) \quad (3.16)$$

where $\bar{A}(T_2)$ is the complex conjugate of $A(T_2)$; $\omega_n = r_n^2\sqrt{EI/ml^4}$, r_n is the nth root of the characteristic equation $1 + \cos(r)\cosh(r) = 0$; and $\Phi_n(s)$ denotes the normalized shape of the nth undamped linear vibration mode, which is given by the following expression:

$$\Phi_n(s) = \frac{1}{\sqrt{l}}\left(\cosh\frac{r_n s}{l} - \cos\frac{r_n s}{l} + \frac{\cos r_n + \cosh r_n}{\sin r_n + \sinh r_n}\left(\sin\frac{r_n s}{l} - \sinh\frac{r_n s}{l}\right)\right)$$

For large n, the numerical evaluation of the r_n and $\Phi_n(s)$ requires retention of an increasingly large number of significant digits because of the presence of hyperbolic (exponential) functions. To avoid this problem, Dowell (1984) and Dugundji (1988) derived simple expressions for higher vibration modes of uniform Euler-Bernoulli beams. McDaniel et al. (2002) proposed a method for estimating the natural frequencies and mode shapes of a multiple-degree-of-freedom system from its forced response vectors.

Substituting Eq. (3.16) into Eq. (3.13) yields

$$\begin{aligned}
D_0^2 v_3 + \frac{EI}{m}v_3^{iv} &= -2i\omega_n \Phi_n D_2 A e^{i\omega_n T_0} - 2i\mu\omega_n \Phi_n A e^{i\omega_n T_0} + \frac{1}{2}\hat{f}e^{i\Omega T_0} \\
&\quad - \frac{EI}{m}\left[\Phi_n'\left(\Phi_n'\Phi_n''\right)'\right]'\left(A^3 e^{3i\omega_n T_0} + 3A^2\bar{A}e^{i\omega_n T_0}\right) \\
&\quad + 2\omega_n^2\left(\Phi_n'\int_l^s\int_0^s \Phi_n'^2 ds ds\right)'\left(A^3 e^{3i\omega_n T_0} + A^2\bar{A}e^{i\omega_n T_0}\right) + cc \\
&\quad - c\omega_n^2\Phi_n\left(iAe^{i\omega_n T_0} - i\bar{A}e^{-i\omega_n T_0}\right)|\Phi_n\left(iAe^{i\omega_n T_0} - i\bar{A}e^{-i\omega_n T_0}\right)|
\end{aligned} \quad (3.17)$$

Here we restrict our discussion to the case of primary resonance of the nth mode (i.e., $\Omega \approx \omega_n$). To express the nearness of this resonance, we introduce the detuning parameter σ defined by $\Omega = \omega_n + \epsilon^2\sigma$. Since the homogeneous problem associated with Eqs. (3.17), (3.14), and (3.15) has a nontrivial solution, the nonhomogeneous problem has a solution only if the right-hand side of Eq. (3.17) is orthogonal to every solution of the adjoint homogeneous problem (Nayfeh, 1981). Therefore, demanding that the right-hand side of Eq. (3.17) be orthogonal to $\Phi_n(s)\exp(-i\omega_n T_0)$, we obtain

$$-2i\omega_n\left(D_2 A + \mu A + \alpha_d c A \mid A \mid\right) - 2\alpha A^2 \bar{A} + \frac{1}{2}fe^{i\sigma T_2} = 0 \quad (3.18)$$

where

$$f = \hat{f} \int_0^l \Phi_n(s)\,ds$$
$$\alpha_d = \frac{8\omega_n}{3\pi} \int_0^l \Phi_n(s)^2 |\Phi_n(s)|\,ds$$
$$\alpha_g = \frac{3EI}{m} \int_0^l \Phi_n'(s)^2 \Phi_n''(s)^2\,ds$$
$$\alpha_i = -\omega_n^2 \int_0^l \left(\int_0^s \Phi_n'(s)^2\,ds \right)^2 ds$$
$$\alpha = \alpha_g + \alpha_i \qquad (3.19)$$

We note that α is the sum of the geometric (hardening) nonlinearity α_g and inertia (softening) nonlinearity α_i and thus denotes the effective nonlinearity corresponding to the nth mode. Also, α is not dimensionless, but rather has dimensions $1/ms^2$.

Substituting the polar form

$$A = \frac{1}{2} a e^{i(\sigma T_2 - \gamma)} \qquad (3.20)$$

into Eq. (3.18), multiplying the result by $\exp\left[i\left(\gamma - \sigma T_2\right)\right]$, and separating real and imaginary parts, we obtain the following autonomous modulation equations:

$$a' = -\mu a - \frac{1}{2}\alpha_d c a^2 + \frac{f}{2\omega_n}\sin\gamma \qquad (3.21)$$
$$a\gamma' = \sigma a - \frac{\alpha}{4\omega_n} a^3 + \frac{f}{2\omega_n}\cos\gamma \qquad (3.22)$$

where the prime indicates differentiation with respect to T_2. Substituting Eq. (3.20) into Eq. (3.16) and then substituting Eq. (3.16) into Eq. (3.7), we find that the beam response is given by

$$v(s,t;\epsilon) = \epsilon a(t)\cos(\Omega t - \gamma)\Phi_n(s) + \cdots \qquad (3.23)$$

3.1.3 Frequency-Response and Force-Response Equations

Periodic solutions of the beam correspond to the fixed points of Eqs. (3.21) and (3.22). To determine these fixed points, we set the right-hand sides of Eqs. (3.21) and (3.22) equal to zero. Now, these two equations can be used to obtain the frequency- and force-response diagrams. The frequency-response diagram is obtained by keeping the forcing amplitude constant while varying the excitation frequency.

In contrast, the force-response diagram is obtained by varying the forcing amplitude while keeping the excitation frequency constant. In both cases, the displacement amplitude of $v(s,t)$ is plotted versus the control parameter (either Ω or a_b).

We use the following two equations, which were derived by setting the right-hand sides of Eqs. (3.21) and (3.22) equal to zero, to obtain the frequency-response and force-response diagrams, respectively:

$$\sigma_{1,2} = \frac{\alpha}{4\omega_n}a^2 \mp \sqrt{\frac{f^2}{4\omega_n^2 a^2} - \left(\mu + \frac{1}{2}\alpha_d c a\right)^2} \qquad (3.24)$$

$$f = 2\omega_n a \sqrt{\left(\mu + \frac{1}{2}\alpha_d c a\right)^2 + \left(\sigma - \frac{\alpha}{4\omega_n}a^2\right)^2} \qquad (3.25)$$

where the subscript 1 and the minus sign refer to the left branch of the frequency-response curve, while the subscript 2 and the plus sign refer to the right branch.

3.2 Experimental Procedure

We excited a steel beam with the dimensions 19.085" $\times \frac{1}{2}$" $\times \frac{1}{32}$" by a base excitation. The density and Young's modulus of the beam were taken as 7810 kg/m^3 and 207 GPa, respectively. The beam was mounted vertically on a steel clamping fixture attached to a MB Dynamics 445 N (100-lb) electrodynamic shaker. The output of the shaker was measured using a PCB 308B02 accelerometer placed on the clamping fixture, and the response of the cantilever beam was measured with a 350 Ohm strain gage mounted approximately 33 mm from the fixed end of the beam. The strain gage formed one arm of a quarter bridge circuit, and its signal was conditioned using a Measurements Group 2310 signal conditioning amplifier. The accelerometer signal was conditioned with a PCB 482A10 amplifier. The accelerometer amplifier and strain gage conditioner were attached in parallel to a Hewlett-Packard 35670A dynamic signal analyzer, which was also used to drive the MB Dynamics SS250 shaker amplifier.

The experiment included four testing sequences related to the third mode and three sequences related to the fourth mode. Each of these testing sequences was run on a separate day. In five of these testing sequences, the frequency was swept while the excitation amplitude was held constant, though the excitation amplitude itself was different for each sequence. In the other two testing sequences, the excitation amplitude was varied while the excitation frequency was held constant. We waited for a long time to ensure steady state before taking any measurement.

3.2.1 Linear Natural Frequencies

The natural frequencies of the beam were determined using the frequency-response function of the signal analyzer. The beam was excited by a 50% burst-chirp low-amplitude excitation, and a uniform window was used to analyze the power spectra of the accelerometer and strain-gage signals. Peaks in the amplitude portion of the frequency-response function give the linear natural frequencies of the beam. To increase confidence in the experimentally obtained linear natural frequencies, we measured the frequency-response functions at several low excitation levels. No noticeable shifts in the peaks were observed. In addition, we made sure that the coherence was close to unity at the corresponding peaks. Also, a periodic checking of the natural frequencies of the beam was done to detect any fatigue damage.

Table 3.1: Experimentally determined third-mode natural frequency.

a_b (m/s^2)	ω_3 (Hz)
$0.10g$	49.078
$0.15g$	49.094
$0.20g$	49.094

Before the beginning of each testing sequence, we measured the natural frequency of the corresponding mode (either third or fourth). In Table 3.1, the measured values of the third-mode natural frequency are listed alongside the value of the base acceleration of the corresponding testing sequence. In the a_b column, g refers to the acceleration due to gravity and has a value equal to 9.8 m/s^2. We note that the value of the estimated third-mode natural frequency is not constant. The minor variation could be explained by the fact that these measurements were done on separate days, and the difference is equal to the frequency resolution of the signal analyzer used to make the measurements. The experimentally measured value of the fourth-mode natural frequency is 96.117 Hz.

3.2.2 Determination of the Beam Displacement

The strain gage essentially measures the strain at the location where it is mounted on the beam. To measure the beam displacement at that point, we need to convert the strain into displacement. In this section, we describe the procedure to convert the strain-gage reading into displacement.

From beam theory, we know that the strain experienced by the strain gage is given by

$$e = y\rho \tag{3.26}$$

where e denotes the strain, ρ denotes the curvature at the location of the strain gage, and y denotes the distance of the strain gage from the beam's neutral axis; that is, $y = \frac{1}{2}b$ where b is the beam thickness. Let $w(s)$ denote the steady-state transverse-displacement amplitude of a point a distance s from the fixed end of the beam. Then, from Eq. (3.23), we can write

$$w(s) = \epsilon\, a\, \Phi_n(s) \tag{3.27}$$

The curvature at a distance s from the fixed end is given by

$$\rho = \frac{\partial^2 w}{\partial s^2}\left[1 + \left(\frac{\partial w}{\partial s}\right)^2\right]^{-\frac{3}{2}} \tag{3.28}$$

For the displacement amplitudes observed in the testing sequences, it was found that the nonlinear expression for the curvature was not necessary. The linear part by itself determines the displacement amplitude to a sufficient degree of accuracy. Hence, we use the following linear expression for the curvature:

$$\rho \approx \frac{\partial^2 w}{\partial s^2} \tag{3.29}$$

which is the commonly used expression for curvature in any strength of materials textbook (Timoshenko and Young, 1968).

Let l_{sg} denote the distance of the strain gage center from the fixed end. Using Eq. (3.27) in Eq. (3.29), we obtain

$$\rho|_{s=l_{sg}} = \epsilon\, a\, \Phi_n''(l_{sg}) \tag{3.30}$$

Alternatively, we can determine the strain from the strain-gage reading as follows:

$$e = \frac{4V_{out}}{V_{excite} G K_g} \tag{3.31}$$

where V_{out} denotes the strain-gage amplifier output in volts, V_{excite} denotes the bridge excitation voltage in volts, G denotes the gain of the strain-gage signal conditioner, and K_g denotes the gage factor of the strain gage. Substitituting Eqs. (3.30) and (3.31) into Eq. (3.26), we obtain

$$a = \frac{8}{\epsilon\, V_{excite} G K_g b\, \Phi_n''(l_{sg})} V_{out}$$

which when substituted into Eq. (3.27) gives

$$w(s) = \frac{8\Phi_n(s)}{V_{excite} G K_g b \, \Phi_n''(l_{sg})} V_{out} \qquad (3.32)$$

Therefore, we only need to multiply the strain-gage reading V_{out} by a constant to obtain the displacement amplitude w at a given s.

Table 3.2: Some constants and their values.

Constant	V_{excite}	G	K_g	b	l_{sg}
Value	10 V	1000	2.095	0.794 mm	32.56 mm

The values of the constants appearing in Eq. (3.32) are listed in Table 3.2. We use w_l to denote the displacement amplitude of the beam tip where $s = l$.

3.3 Parameter Estimation Procedure

We estimate the parameters (ζ, $\bar{c}$, α) describing the weakly damped, weakly nonlinear beam system from the experimental frequency-response results. We know that, for a given excitation level, the amplitude at the peak of the corresponding frequency-response curve depends on the damping value (Nayfeh and Mook, 1979). And the effect of the nonlinearity is essentially to shift the peak away from the natural frequency ω_n. For a system with hardening nonlinearity, the peak is shifted to the right; and in the case of a softening nonlinearity it is shifted to the left. The magnitude of the shift depends on the strength of the nonlinearity. Thus, knowing the amplitude at the peak and the frequency shift, it is possible to estimate the damping coefficient(s) and the effective nonlinearity of a system. The detailed estimation procedure is described in the following subsections.

Table 3.3 lists the coordinates of the peaks of the experimentally obtained third-mode frequency-response curves for different base acceleration levels. The peak displacement amplitude of the beam tip is denoted by w_l^*, and Ω^* is used to denote the value of the excitation frequency at the peak. We note that ω_3, given in Table 3.1, denotes the linear resonance frequency, whereas Ω^* (in Table 3.3) denotes the nonlinear resonance frequency.

Table 3.3: Coordinates of the peak of the frequency-response curve.

a_b (m/s^2)	Ω^* (Hz)	w_l^* (mm)
$0.10g$	48.963	2.415
$0.15g$	48.917	3.220
$0.20g$	48.885	3.876

3.3.1 Estimation of the Damping Coefficients

Linear Damping Model

We can easily estimate the linear viscous damping coefficient μ from the experimental frequency-response curve. For the linear damping model ($c = 0$), it follows from Eq. (3.24) that the peak of the frequency-response curve corresponds to

$$\mu = \frac{f}{2\omega_n a^*} \tag{3.33}$$

where a^* denotes the value of a at the peak. So, by measuring a^* and knowing the values of f and ω_n, we can estimate the linear damping coefficient. Using the definitions of μ, f, and $w(s)$, we rewrite Eq. (3.33) in terms of physical quantities as

$$\zeta = \frac{a_b \xi_n \Phi_n(l)}{2\omega_n^2 w_l^*} \times 10^3 \tag{3.34}$$

where $\xi_n = \int_0^l \Phi_n(s)ds$, ω_n is the nth natural frequency in rad/sec, and w_l^* denotes the peak displacement amplitude of the beam tip in mm.

Nonlinear Damping Model

With the addition of quadratic damping (air drag), we complicate our model. But the advantage, as we would see later, is that the experimental and theoretical frequency- and force-response curves are in much better agreement when compared to the linearly damped case.

At the peak of the frequency-response curve, it follows from Eq. (3.24) that

$$\mu + \frac{1}{2}\alpha_d c a^* = \frac{f}{2\omega_n a^*} \tag{3.35}$$

where a^* denotes the value of a at the peak. But now we have two unknowns (μ and c) and just one equation. We, therefore, use results of two testing sequences to determine the damping coefficients c and μ. Using Eq. (3.35) for two testing sequences, denoted by subscripts 1 and 2, we have

$$c = \frac{1}{\alpha_d(a_2^* - a_1^*)}\left(\frac{f_2}{\omega_n a_2^*} - \frac{f_1}{\omega_n a_1^*}\right)$$

$$\mu = \frac{f_2}{2\omega_n a_2^*} - \frac{1}{2}\alpha_d c\, a_2^*$$

In terms of physical quantities, we have

$$\bar{c} = \frac{\xi_n \Phi_n(l)^2}{\alpha_d(w_{l2}^* - w_{l1}^*)}\left[\frac{a_{b_2}}{\omega_n w_{l2}^*} - \frac{a_{b_1}}{\omega_n w_{l1}^*}\right] \times 10^6 \qquad (3.36)$$

$$\zeta = \frac{a_{b_2}\xi_n \Phi_n(l)}{2\omega_n^2 w_{l2}^*} \times 10^3 - \frac{\alpha_d w_{l2}^* \bar{c}}{2\omega_n \Phi_n(l)} \times 10^{-3} \qquad (3.37)$$

where w_{l1}^* and w_{l2}^* denote the peak displacement amplitudes of the beam tip (in mm) corresponding to the testing sequences 1 and 2, respectively.

3.3.2 Nonlinearity Estimation

At the peak of the frequency-response curve, it follows from Eqs. (3.24), (3.33), and (3.35) that

$$\alpha = \frac{4\omega_n \sigma^*}{a^{*2}}$$

where a^* and σ^* denote the values of a and σ, respectively, at the peak. In terms of physical quantities, we have

$$\alpha = \frac{4\omega_n \Phi_n(l)^2(\Omega^* - \omega_n)}{w_l^{*2}} \times 10^6 \qquad (3.38)$$

where w_l^* and Ω^* denote the tip displacement amplitude (in mm) and excitation frequency (in rad/sec) at the peak, respectively. We note that Eq. (3.38) is applicable to both the linear as well as the nonlinear damping models.

3.3.3 Curve-Fitting the Frequency-Response Data

The parameter values can also be estimated by curve-fitting the experimental frequency-response data points. This is the approach used by Krauss and Nayfeh (1999a,b) to estimate nonlinear parameters of a system. We used the MATLAB function `nlinfit`, which performs a nonlinear least-squares data fitting using a Gauss-Newton method, to fit a theoretical model to the experimental data points. More specifically, we estimated the parameters in Eq. (3.24) by fitting it to the experimental frequency-response data set. A drawback of the curve-fitting method is its sensitivity to and dependence on the initial guesses. In the next section, we compare the results obtained, for the fourth mode, using the proposed estimation technique with those obtained by the curve-fitting method.

3.3.4 Fixing f and ω_n

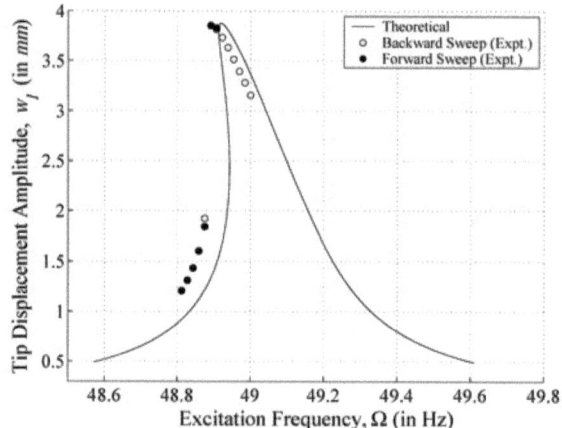

Figure 3.1: Experimentally and theoretically obtained third-mode frequency-response curves for $a_b = 0.2g$ and $\omega_3 = 49.094$ Hz using the linear damping model.

There were some differences between the experimentally and theoretically obtained frequency-response curves. In Fig. 3.1, we have plotted an experimentally observed frequency-response curve along with the theoretical one for the third mode. We used the theoretical value of α, the estimated value of μ (and c), and the experimentally determined values of f and ω_3 in Eq. (3.24) to obtain the theoretical frequency-response curve. The value of f was determined from the accelerometer reading, and ω_3 was determined using the frequency-response function of the signal analyzer. We observe in Fig. 3.1 that the theoretically obtained frequency-response curve is to the right of the experimental one. This

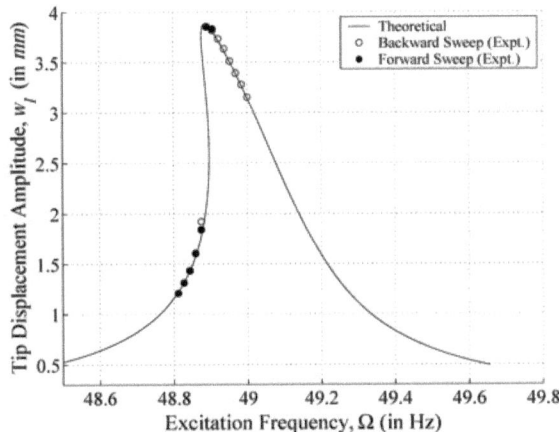

Figure 3.2: Experimentally and theoretically obtained third-mode frequency-response curves for $a_b = 0.23g$ and $\omega_3 = 49.06$ Hz using the linear damping model.

indicates that the value of ω_3 used in Eq. (3.24) needs to be lowered. Also, we observe that the spike in the experimental frequency-response curve is wider than the theoretical one. This indicates that the value of f used in Eq. (3.24) needs to be increased.

For better agreement between the third-mode experimental and theoretical frequency-response results, we had to lower the value of ω_3 by nearly 0.05 Hz, which is around 0.1% of the measured third natural frequency, and we had to increase the value of f by nearly 15% for the linear damping model and by 8% for the nonlinear damping model. The probable reason for the shift in the frequency-response curve could be the non-inclusion of gravity in the equation of motion and/or of higher-order terms in the modulation equations. But on investigation later, we found that inclusion of higher-order terms did not affect the results much. Also, with the addition of gravity, which tends to lower the natural frequencies, the problem still persisted. We believe the assumption of ideal-clamp boundary condition could be contributing to the shift in the results. Tabaddor (2000) studied the influence of nonideal-clamp boundary conditions and found significant changes in the model behavior. This needs to be studied further in the present context. On the other hand, we believe the difference in the value of f is probably due to improper calibration of the accelerometer and/or due to an error in the measurement of the strain-gage location from the fixed end of the beam. On investigation, it was later found that the accelerometer was reading a value lower than the actual base acceleration value.

Figure 3.2 compares again the experimental and theoretical frequency-response curves with the theoretical curve obtained using the modified values of f and ω_3. The two are in much better agreement now. We also note that, without the modifications to the values of f and ω_3, the estimations of the parameters (ζ, $\bar{c}$, α) were also not correct.

Table 3.4: Modified values of ω_3. The superscripts l and n refer to the linear and nonlinear damping models, respectively.

a_b (m/s^2)	ω_3^l (Hz)	ω_3^n (Hz)
$0.10g$	49.035	49.035
$0.15g$	49.040	49.034
$0.20g$	49.060	49.055

The modified values of ω_3 for each of the base excitation levels are listed in Table 3.4. They were determined by lowering the value of ω_3 used in Eq. (3.24) till the theoretical frequency-response curve got closer to the experimental one. The values of a_b, for all of the third-mode testing sequences, were increased by 15% in the case of the linear damping model and by 8% in the case of the nonlinear damping model.

3.3.5 Critical Forcing Amplitude

The proposed estimation technique relies solely on the coordinates of the peak of the frequency-response curve and hence, for better results, the peak location has to be determined accurately. The coordinates of the peaks of the frequency-response curves are obtained through a cubic spline interpolation of the experimental data points. But if there is a jump in the frequency-response diagram, then it would be difficult to obtain a good cubic spline interpolant and hence, in such a case, the peak location cannot be measured accurately. We, therefore, suggest that forcing levels which do not lead to jumps in the frequency-response diagram be used.

Let f_{cr} denote the critical value of the forcing amplitude f marking the boundary between the values of f leading to jumps and those not leading to jumps. For the case of linear damping ($c = 0$), f_{cr} is given by

$$f_{cr} = 8\mu\omega_n \sqrt{\frac{2\mu\omega_n}{3\sqrt{3}|\alpha|}}$$

which can be rewritten in terms of physical quantities as follows:

$$a_{b_{cr}} = \frac{8\zeta\omega_n^3}{3\xi_n}\sqrt{\frac{2\sqrt{3}\zeta}{|\alpha|}}$$

where $a_{b_{cr}}$ denotes the critical base acceleration. A detailed derivation procedure of the critical forcing amplitude is presented in Chapter 4. To determine $a_{b_{cr}}$, we use the theoretical value of α (Eq. (3.19)) and the experimental values of ζ and ω_n in the above equation. Here, we reiterate that, to obtain good results, a base excitation amplitude $a_b < a_{b_{cr}}$ be used.

3.4 Results

In this section, we first present the estimates of the parameters (ζ, $\bar{c}$, α) obtained using the proposed estimation technique. To validate the proposed estimation technique, we compare the experimentally and theoretically obtained third-mode frequency- and force-response curves. The theoretical frequency- and force-response curves are obtained using the estimated parameter values in Eqs. (3.24) and (3.25), respectively. Theoretical results for both the linear as well as the nonlinear damping models are presented. To further boost the confidence in the proposed estimation technique, we compare its fourth-mode frequency- and force-response curves with those obtained using the curve-fitting method.

3.4.1 Third-Mode Estimation Results

For the linear damping model, the value of the linear viscous damping factor ζ is estimated using Eq. (3.34). The estimated values of the damping factor ζ obtained using the experimental results of the third-mode frequency-response testing sequences are listed in Table 3.5. We note that the value of ζ is not constant but shows a variation of approximately 12%. This is expected as damping is essentially a complex phenomenon, which depends upon the frequency and amplitude of vibration.

For the nonlinear damping model, the values of the damping coefficients ζ and $\bar{c}$ are estimated using Eqs. (3.36) and (3.37). The values of the linear viscous damping factor ζ and the quadratic

Table 3.5: Estimated values of the third-mode viscous damping factor ζ of the linear damping model.

a_b (m/s^2)	ζ
$0.10g$	1.251×10^{-3}
$0.15g$	1.407×10^{-3}
$0.20g$	1.557×10^{-3}

Table 3.6: Estimated values of the third-mode damping coefficients ζ and $\bar{c}$ of the nonlinear damping model.

a_b (m/s^2)	ζ	$\bar{c}$ $(1/m)$
$0.10g$	7.341×10^{-4}	0.677
$0.15g$	6.975×10^{-4}	0.732
$0.20g$	6.262×10^{-4}	0.801

damping coefficient $\bar{c}$ obtained using the experimental results of the third-mode frequency-response testing sequences are listed in Table 3.6. We note that the values of ζ and $\bar{c}$ are not constant but show a variation of around 7%.

Table 3.7: Estimated values of the third-mode effective nonlinearity α. The superscripts l and n refer to the linear and nonlinear damping models, respectively.

a_b (m/s^2)	α^l $(1/ms^2)$	α^n $(1/ms^2)$
$0.10g$	-7.888×10^8	-7.888×10^8
$0.15g$	-7.581×10^8	-7.210×10^8
$0.20g$	-7.447×10^8	-7.233×10^8

For both of the linear as well as the nonlinear damping models, the value of the effective nonlinearity α is estimated using Eq. (3.38). Table 3.7 lists the estimated values of α for each of the third-mode frequency-response testing sequences. There is a slight variation in the estimated values of α, but all of them are close to the theoretical value of $\alpha = -7.543 \times 10^8$ obtained using Eq. (3.19). The difference in the values of α^l and α^n is due to the difference in the values of ω_3^l and ω_3^n (refer to Table 3.4).

Chapter 3. Parametric System Identification   53

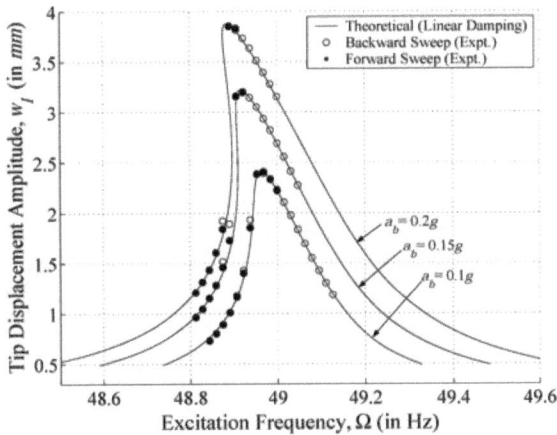

Figure 3.3: Experimentally and theoretically obtained third-mode frequency-response curves for $a_b = 0.1g$, $0.15g$, and $0.2g$ using the linear damping model.

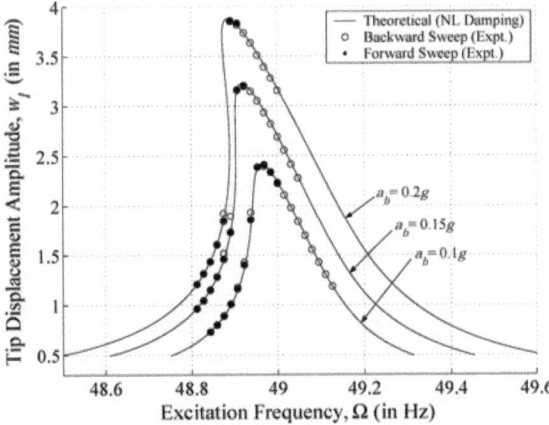

Figure 3.4: Experimentally and theoretically obtained third-mode frequency-response curves for $a_b = 0.1g$, $0.15g$, and $0.2g$ using the nonlinear damping model.

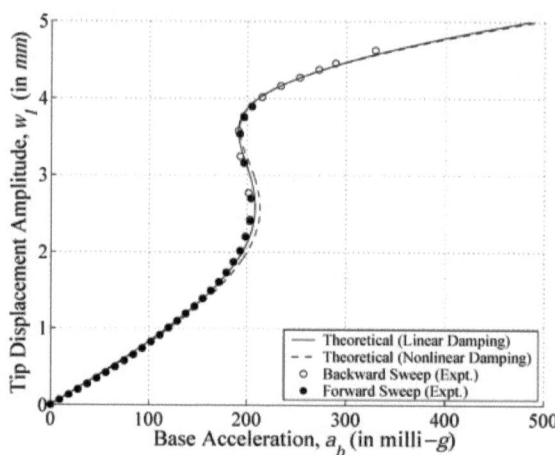

Figure 3.5: Experimentally and theoretically obtained third-mode force-response curves using the linear and nonlinear damping models for $\Omega = 48.891$ Hz.

Next, we compare the experimentally and theoretically obtained third-mode frequency- and force-response curves. The theoretical curves are obtained using Eqs. (3.24) and (3.25). For parameters appearing in these equations, corresponding values estimated from the experimental results are used. Figure 3.3 compares the experimentally and theoretically obtained third-mode frequency-response curves using the linear damping model. The parameter values shown in Tables 3.5 and 3.7 were used to obtain the theoretical curves. Figure 3.4 compares the experimentally and theoretically obtained third-mode frequency-response curves using the nonlinear damping model. The parameter values shown in Tables 3.6 and 3.7 were used to obtain the theoretical curves. Figure 3.5 compares the experimental third-mode force-response curve with those obtained theoretically using the linear and nonlinear damping models. The agreement between the experimental results and the results obtained using the theoretical models is very good. Especially, the nonlinear damping model does a very good job predicting values very close to all of the experimental data points. Also, in the case of the linear damping model, if a value of ζ estimated for one particular excitation amplitude is used in plotting the theoretical frequency-response curve of another excitation amplitude, then it leads to an overshoot or undershoot of the peak of the curve. This is not the case with the nonlinear damping model. Any pair

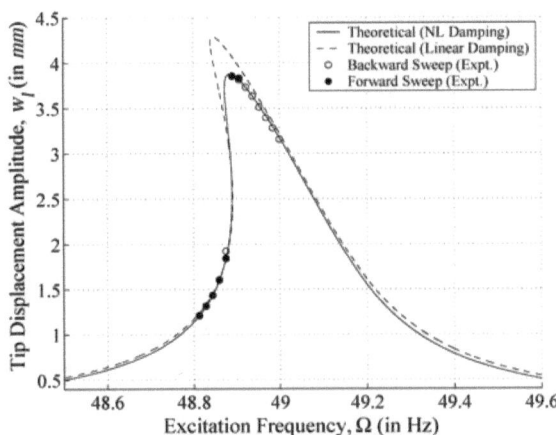

Figure 3.6: Overshoot in the peak of the third-mode frequency-response curve obtained using the linear damping model.

of (ζ, $\bar{c}$) values from Table 3.6 can be used to correctly predict the response curves of other testing sequences. This behavior is shown in Fig. 3.6, where the estimated values of ζ (and $\bar{c}$) from the second testing sequence ($a_b = 0.15g$) were used to plot the theoretical frequency-response curve of the third testing sequence ($a_b = 0.2g$). We observe an overshoot using the linear damping model, whereas the results obtained using the nonlinear damping model match very well the corresponding experimental results.

3.4.2 Comparison with Curve-Fitting Method

To further validate the proposed estimation technique, we compare the fourth-mode frequency- and force-response curves obtained using the proposed technique with those obtained using the curve-fitting method.

Tables 3.8 and 3.9 list the estimated values of the parameters of the linear and nonlinear damping models obtained using the proposed estimation technique and the curve-fitting method. Using these estimated values, we plotted the frequency- and force-response curves for the linear and nonlinear

Table 3.8: Comparison of the estimates of the damping factor ζ and the effective nonlinearity α for the fourth mode using the linear damping model. The superscripts p and cf refer to the proposed estimation technique and the curve-fitting method, respectively.

a_b (m/s²)	ζ^p	α^p (1/ms²)	ζ^{cf}	α^{cf} (1/ms²)
$0.075g$	6.951×10^{-4}	-1.621×10^{10}	6.955×10^{-4}	-1.766×10^{10}
$0.100g$	7.220×10^{-4}	-1.374×10^{10}	7.220×10^{-4}	-1.430×10^{10}

Table 3.9: Comparison of the estimates of the fourth-mode damping coefficients ζ and $\bar{c}$ and the effective nonlinearity α using the nonlinear damping model. The superscripts p and cf refer to the proposed estimation technique and the curve-fitting method, respectively.

a_b (m/s²)	ζ^p	$\bar{c}^p$ (1/m)	α^p (1/ms²)	ζ^{cf}	$\bar{c}^{cf}$ (1/m)	α^{cf} (1/ms²)
$0.075g$	5.810e-4	0.624	-1.621e10	6.157e-4	0.471	-1.751e10
$0.100g$	5.810e-4	0.624	-1.300e10	6.083e-4	0.501	-1.325e10

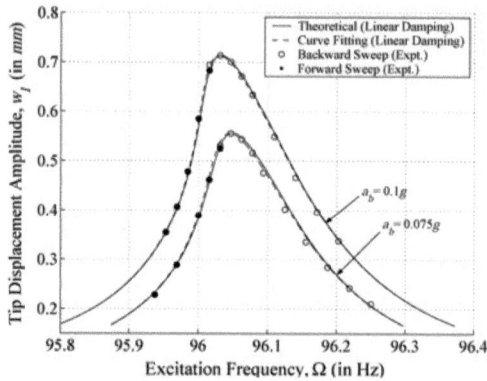

Figure 3.7: Comparison of the fourth-mode frequency-response curves obtained using the proposed technique and the curve-fitting method for $a_b = 0.075g$ and $0.1g$ using the linear damping model.

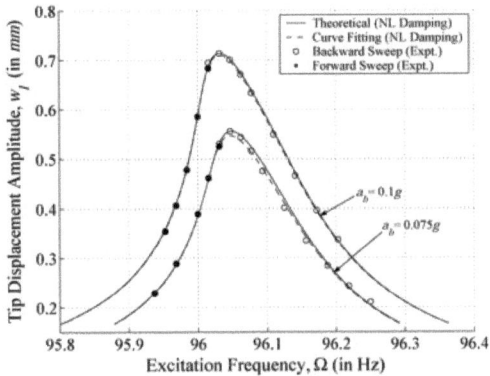

Figure 3.8: Comparison of the fourth-mode frequency-response curves obtained using the proposed technique and the curve-fitting method for $a_b = 0.075g$ and $0.1g$ using the nonlinear damping model. Note: In the legend, NL stands for Nonlinear.

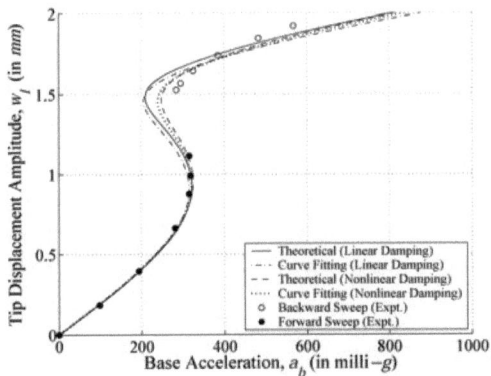

Figure 3.9: Comparison of the fourth-mode force-response curves obtained using the proposed technique and the curve-fitting method for $\Omega = 95.844$ Hz using the linear and nonlinear damping models.

damping models, which are displayed in Figs. 3.7-3.9. The agreement between the results obtained with the two different techniques is very good.

The linear damping model was found to have a serious drawback. The value of the linear viscous damping factor ζ was also determined experimentally using the half-power method and was found to be $\zeta = 6.569 \times 10^{-4}$. But the values of ζ estimated using the linear damping model are higher than the measured experimental value. On the other hand, the values of ζ estimated by the nonlinear damping model are less than the measured experimental value, as expected.

3.5 Closure

A simple and straightforward parametric identification procedure for estimating the nonlinear parameters describing a single-mode response of a weakly nonlinear cantilever beam is presented. Using information of the peak locations of one or two (depending on the damping model) frequency-response curves, one can estimate to a sufficient degree of accuracy the parameters of the nonlinear model describing the cantilever beam system. This method is applicable to any SDOF weakly nonlinear system with cubic geometric and inertia nonlinearities. However, we note that the proposed method cannot be used to estimate the individual geometric and inertia nonlinearity contributions.

The results obtained using the linear and nonlinear damping models are qualitatively similar but quantitatively different. For the linear viscous damping factor, the linear damping model estimated a value much higher than the one determined experimentally using the half-power method. Also, the theoretical frequency-response curve obtained using the linear damping model does not pass through all of the experimental data points. This shows that a linear damping model does not model the beam system well. It is reasonable to assume that large deflections of a blunt body like the beam would give rise to significant air damping, which is proportional to the square of the velocity. So, inclusion of the quadratic damping term seems physically justified. This justification was strengthened by the fact that the nonlinear damping model with a quadratic damping term was able to predict results close to the experimental data points; it also estimated for the linear viscous damping factor a value less than the measured experimental value, as expected.

The estimated value of the effective nonlinearity using the proposed estimation technique is close to the theoretical value; it also leads to a good agreement between the experimentally and theoretically obtained force-response curves. Results obtained using the proposed technique are similar to those obtained by the curve-fitting method.

The classic backbone curve method requires determination of peak locations of multiple frequency-response curves corresponding to different forcing levels and hence is time consuming. On the other hand, the proposed method is simple and also demonstrates that the effective nonlinearity can be determined accurately from the peak location of a single frequency-response curve. In addition, the proposed method is more direct and straightforward and does not involve any least-squares curve fitting. Finally, the new estimation technique is also very robust, which is demonstrated by the fact that it led to a very good agreement between the experimentally and theoretically obtained frequency- and force-response curves for both the third mode as well as the fourth mode.

Chapter 4

Determination of Jump Frequencies

It is a well-known fact that the nonlinearity present in a system leads to jumps in the frequency- and force-response curves (Nayfeh and Mook, 1979). As shown in Fig. 4.1, the frequency-response curve of a Duffing oscillator is bent either to the left or to the right, depending on whether the type of the nonlinearity is softening or hardening. The bending of the frequency-response curve leads to a jump in the response amplitude when the excitation frequency is swept from left-to-right or right-to-left. The response amplitude increases at a jump-up point and decreases at a jump-down point. Between the jump points, multiple solutions exist for a given value of the excitation frequency, and the initial conditions determine which of these solutions represents the actual response of the system. The jump points of a frequency-response curve coincide with the turning points of the curve where saddle-node bifurcations occur. The goal of this chapter is to determine the minimum forcing amplitude that would lead to jumps in the frequency-response curves of single-degree-of-freedom (SDOF) nonlinear systems and to also locate the jump-up and jump-down points in the frequency-response curve when the forcing amplitude is above the minimum value.

Friswell and Penny (1994) and Worden (1996) computed the bifurcation points of the frequency-response curve of a Duffing oscillator with linear damping. They used the method of harmonic balance to obtain the frequency-response function. To compute the jump frequencies, Worden (1996) set the discriminant of the frequency-response function, which is a cubic polynomial in the square of the amplitude, equal to zero, while Friswell and Penny (1994) used a numerical approach based on Newton's method. Their first-order approximation results agree well with the "exact" results. But

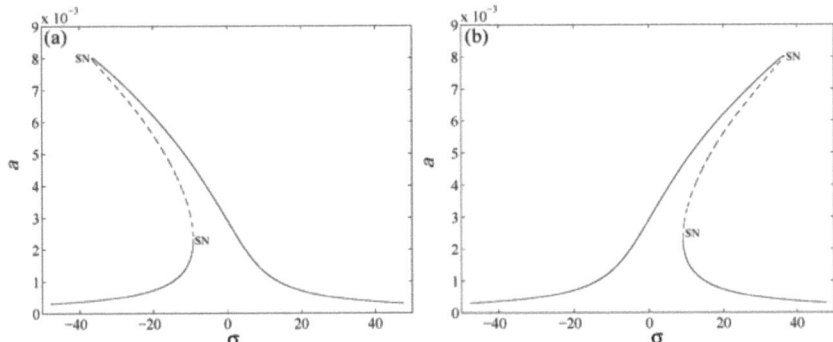

Figure 4.1: Typical frequency-response curves of a Duffing oscillator with (a) softening nonlinearity and (b) hardening nonlinearity. Dashed lines (- -) indicate unstable solutions and SN refers to a saddle-node bifurcation.

for systems with higher-order geometric, inertia, and/or damping nonlinearities, a more general and simple method of determining the jump frequencies is required. In this chapter, we present two methods based on the elimination theory of polynomials (Griffiths, 1947; Wee and Goldman, 1995a), which can be used to determine both the critical forcing amplitude as well as the jump frequencies in the case of SDOF nonlinear systems. Also, the methods are devoid of convergence problems associated with bad initial guesses and have the potential of being applicable to multiple-degree-of-freedom (MDOF) nonlinear systems (Wee and Goldman, 1995b; Cox, Little, and O'Shea, 1997). The proposed methods are outlined in the context of a single-mode response of an externally excited cantilever beam possessing cubic geometric and inertia nonlinearities and linear and quadratic damping.

4.1 Theory

4.1.1 Frequency-Response Function

As the cantilever beam constitutes a weakly damped, weakly nonlinear system, we use the method of multiple scales (Nayfeh, 1981) to derive the modulation equations governing the amplitude and phase of the excited mode of the cantilever beam. In the process of deriving the modulation equations, we define the following quantities:

$$\mu \equiv \zeta \omega_n, \quad \sigma \equiv \Omega - \omega_n, \quad f \equiv a_b \int_0^l \Phi_n\, ds, \quad c \equiv \frac{4\omega_n}{3\pi} \bar{c} \int_0^l \Phi_n^2 |\Phi_n|\, ds$$

where ζ is the linear viscous damping factor, ω_n is the nth natural frequency of the beam, Ω is the excitation frequency, a_b is the base acceleration, l is the length of the beam, s is the arclength, $\Phi_n(s)$ is the normalized nth mode shape, and $\bar{c}$ is the quadratic damping coefficient per unit mass and length.

Seeking a first-order uniform expansion of the transverse displacement $v(s,t)$ of the beam, we obtain

$$v(s,t) \approx a(t)\cos(\Omega t - \gamma)\Phi_n(s) + \cdots$$

and the modulation equations governing the amplitude a and phase γ of the response are given by

$$\dot{a} = -\mu a - c a^2 + \frac{f}{2\omega_n}\sin\gamma \tag{4.1}$$

$$a\dot{\gamma} = \sigma a - \frac{\alpha}{4\omega_n}a^3 + \frac{f}{2\omega_n}\cos\gamma \tag{4.2}$$

where α is the effective nonlinearity comprising the contributions of the geometric and inertia nonlinearities, and the overdot indicates differentiation with respect to time t. A detailed description of the derivation of the modulation equations is given in Chapter 3.

Periodic solutions of the beam correspond to the fixed points of Eqs. (4.1) and (4.2). To determine these fixed points, we set the right-hand sides of Eqs. (4.1) and (4.2) equal to zero. We, thus, obtain the following frequency-response function relating the response amplitude a and the excitation frequency Ω (or σ):

$$\sigma_{1,2} = \frac{\alpha}{4\omega_n}a^2 \mp \sqrt{\frac{f^2}{4\omega_n^2 a^2} - (\mu + ca)^2} \tag{4.3}$$

where the subscript 1 and the '−' sign refer to the left branch of the frequency-response curve, while the subscript 2 and the '+' sign refer to the right branch. Equation (4.3) can be rewritten in polynomial form as

$$\mathcal{F}(a,\sigma) = a^6 + pa^4 + qa^3 + ra^2 + s = 0 \tag{4.4}$$

where

$$p = \frac{16\omega_n^2}{\alpha^2}(c^2 - \frac{\alpha}{2\omega_n}\sigma), \quad q = \frac{32\omega_n^2}{\alpha^2}\mu c, \quad r = \frac{16\omega_n^2}{\alpha^2}(\mu^2 + \sigma^2), \quad s = -\frac{4f^2}{\alpha^2}$$

The frequency-response function can also be written as a polynomial function in σ as follows:

$$\mathcal{F}(a,\sigma) = p\sigma^2 + q\sigma + r = 0 \tag{4.5}$$

where

$$p = \frac{16\omega_n^2}{\alpha^2}a^2, \quad q = -\frac{8\omega_n}{\alpha}a^4, \quad r = a^6 - \frac{4f^2}{\alpha^2} + \frac{16\omega_n^2}{\alpha^2}(c^2a^4 + 2c\mu a^3 + \mu^2 a^2)$$

4.1.2 Sylvester Resultant

The resultant of two polynomials is defined as the product of all of the differences between the roots of the polynomials and is a polynomial in the coefficients of the two polynomials (Griffiths, 1947). Consider two polynomials $f(x)$ and $g(x)$ defined as

$$f(x) \equiv \sum_{i=0}^{n} a_i x^i, \quad a_n \neq 0, \quad g(x) \equiv \sum_{i=0}^{m} b_i x^i, \quad b_m \neq 0$$

Then, the Sylvester resultant of $f(x)$ and $g(x)$, denoted by $\mathcal{R}(f,g)$, is given by (Wee and Goldman, 1995a)

$$\mathcal{R}(f,g) = \begin{vmatrix} a_n & a_{n-1} & \cdots & \cdots & a_1 & a_0 & 0 & \cdots & \cdots & 0 \\ 0 & a_n & a_{n-1} & \cdots & \cdots & a_1 & a_0 & 0 & \cdots & 0 \\ \cdots & \cdots & \cdots & & & \cdots & \cdots & \cdots & & \\ 0 & \cdots & 0 & a_n & a_{n-1} & \cdots & \cdots & \cdots & \cdots & a_0 \\ b_m & b_{m-1} & \cdots & b_1 & b_0 & 0 & \cdots & \cdots & \cdots & 0 \\ 0 & b_m & b_{m-1} & \cdots & b_1 & b_0 & 0 & \cdots & \cdots & 0 \\ \cdots & \cdots & \cdots & & & \cdots & \cdots & \cdots & & \\ 0 & \cdots & \cdots & 0 & b_m & b_{m-1} & \cdots & \cdots & \cdots & b_0 \end{vmatrix}$$

A necessary and sufficient condition for $f(x)$ and $g(x)$ to have a common root is that the resultant $\mathcal{R}(f,g)$ be equal to zero (Griffiths, 1947). The discriminant Δ of a polynomial $f(x)$, of order m, is related to the resultant $\mathcal{R}(f,f')$ in the following manner:

$$\mathcal{R}(f,f') = (-1)^{\frac{1}{2}m(m-1)} a_m \Delta$$

where a_m is the coefficient of the x^m term in the polynomial $f(x)$. We know that $f(x) = 0$ has two equal roots iff $f(x) = 0$ and $f'(x) = 0$ have a common root, and hence iff $\mathcal{R}(f,f') = 0$. We use this idea to determine the critical forcing amplitude and jump frequencies.

4.1.3 Critical Forcing Amplitude

For a low excitation amplitude, we do not observe the jump phenomenon and the frequency-response curve is single-valued; that is, for every value of Ω there is a unique value of a. But in the case of a large excitation amplitude, we observe jumps, and for a range of Ω values there exist multiple values of a for a given value of Ω, as seen in Fig. 4.1. Let f_{cr} denote the critical value of f marking the boundary between the values of f leading to jumps and those not leading to jumps. The frequency-response curve for $f = f_{cr}$ has an inflection point, which we denote by (σ_{cr}, a_{cr}), where the frequency-response function $\mathcal{F}(a, \sigma_{cr}) = 0$ has three positive real roots equal to a_{cr}. Therefore, the derivative of the frequency-response function with respect to the response amplitude a, denoted by $\mathcal{F}'(a, \sigma_{cr}) = 0$, has two real roots equal to a_{cr}, which requires that the resultant $\mathcal{R}(\mathcal{F}', \mathcal{F}'')$ be equal to zero at the inflection point (σ_{cr}, a_{cr}). Thus, using Eq. (4.5), we obtain

$$\mathcal{S}(a_{cr}) \equiv \mathcal{R}(\mathcal{F}', \mathcal{F}'')\big|_{a=a_{cr}} = \sum_{i=0}^{6} b_i a_{cr}^i = 0 \qquad (4.6)$$

where

$$b_0 = 144\, c^2 \mu^2 \omega_n^4, \quad b_1 = 384\, c^3 \mu \omega_n^4, \quad b_2 = 64\, \omega_n^2 (\alpha^2 \mu^2 + 4\, c^4 \omega_n^2),$$
$$b_3 = 168\, c\alpha^2 \mu \omega_n^2, \quad b_4 = 96\, c^2 \alpha^2 \omega_n^2, \quad b_5 = 0, \quad b_6 = -3\, \alpha^4$$

We now have a sextic polynomial equation in the response amplitude at the inflection point a_{cr}. Using the resultant, we basically eliminate σ_{cr} and obtain a polynomial equation in a_{cr} only. By using Eq. (4.4), we can eliminate a_{cr} and obtain a polynomial equation in σ_{cr}, but that would involve a more number of computations. Also, in that case spurious solutions appear while solving for σ_{cr}.

Knowing the b_i, one can easily compute the value of a_{cr} numerically. Of the six roots of $\mathcal{S}(a_{cr}) = 0$, only one turns out to be real and positive. Once we know the value of a_{cr}, substituting it into $\mathcal{F}''(a, \sigma_{cr}) = 0$ gives us the critical excitation frequency σ_{cr}. Using the values of σ_{cr} and a_{cr} in Eqs. (4.4) or (4.5), we obtain the critical forcing amplitude f_{cr}.

For the case of linear damping ($c = 0$), a closed-form solution for the critical forcing amplitude is possible. The corresponding expressions of f_{cr}, a_{cr}, and σ_{cr} are as follows:

$$f_{cr} = 8\, \mu \omega_n \sqrt{\frac{2\, \mu \omega_n}{3\sqrt{3}|\alpha|}}, \quad a_{cr} = \sqrt{\frac{8\, \mu \omega_n}{\sqrt{3}|\alpha|}}, \quad \sigma_{cr} = \pm\sqrt{3}\, \mu$$

where the '+' sign is for systems with effective hardening nonlinearity (i.e., $\alpha > 0$), and the '−' sign is for systems with effective softening nonlinearity (i.e., $\alpha < 0$).

4.1.4 Jump Frequencies

For $f > f_{cr}$, we observe jumps in the frequency-response curve, as seen in Fig. 4.1. At the jump points, which we denote by (σ^*, a^*), the frequency-response function $\mathcal{F}(a, \sigma^*) = 0$ has two positive real roots equal to a^*, which requires that the resultant $\mathcal{R}(\mathcal{F}, \mathcal{F}')$ be equal to zero at those points. Using Eq. (4.5), we thus obtain a twelfth-order polynomial equation in a^* as follows:

$$\mathcal{S}(a^*) \equiv \mathcal{R}(\mathcal{F}, \mathcal{F}')\big|_{a=a^*} = \sum_{i=0}^{12} c_i a^i \big|_{a=a^*} = 0 \qquad (4.7)$$

where the c_i are functions of known physical quantities. The values of a^* can be easily computed numerically. Of the twelve roots of $\mathcal{S}(a^*) = 0$, only two turn out to be real and positive. Once we know the value of a^*, substituting it into $\mathcal{F}'(a, \sigma^*) = 0$ gives us the the jump frequency σ^*. But for each value of a^*, we obtain two values of σ^*, one of which is spurious. To pin-point the spurious σ^* solution, we check if $\mathcal{F}(a, \sigma^*) = 0$ leads to two positive real roots equal to a^*. If it does not, then that particular σ^* solution is spurious and is discarded. Alternatively, we could also determine σ^* using Eq. (4.3). The knowledge of the type of nonlinearity can be used to decide whether the jump points lie on the left or the right branch.

4.1.5 Gröbner Basis

A Gröbner basis for the polynomials $\{f_1, f_2, \ldots, f_n\}$ comprises a set of polynomials $\{\mathcal{G}_1, \mathcal{G}_2, \ldots, \mathcal{G}_m\}$ that have the same collection of roots as the original polynomials (Cox et al., 1997). Like the Sylvester resultant, the Gröbner bases also can be used to determine the critical forcing amplitude and jump frequencies. The advantage of using Gröbner bases over resultants is that we do not obtain any spurious solutions while solving for the jump frequencies σ^*. But in general, resultants are more efficient than Gröbner bases.

To determine the critical forcing amplitude, we use the fact that $\mathcal{F}'(a, \sigma) = 0$ and $\mathcal{F}''(a, \sigma) = 0$ at the inflection point (σ_{cr}, a_{cr}). We begin by computing a Gröbner basis for the polynomials $\mathcal{F}'(a, \sigma)$ and $\mathcal{F}''(a, \sigma)$, and thus obtain two polynomials $\mathcal{G}_1$ and $\mathcal{G}_2$, which also vanish at the inflection point (σ_{cr}, a_{cr}) and have a unique structure as we shall see later. Using Eqs. (4.4) or (4.5) and the lex order $\sigma > a$ (i.e., forcing polynomials containing σ appear at a later order compared to polynomials containing only a), we obtain

$$\mathcal{G}_1(a_{cr}) = \sum_{i=0}^{6} b_i a_{cr}^i = 0 \qquad (4.8)$$

$$\mathcal{G}_2(\sigma_{cr}, a_{cr}) = 96\, c a \mu \omega_n^3 \sigma_{cr} + \sum_{i=0}^{5} c_i a_{cr}^i = 0 \qquad (4.9)$$

where

$$b_0 = 144\, c^2 \mu^2 \omega_n^4, \quad b_1 = 384\, c^3 \mu \omega_n^4, \quad b_2 = 64\, \omega_n^2 (\alpha^2 \mu^2 + 4\, c^4 \omega_n^2),$$
$$b_3 = 168\, c \alpha^2 \mu \omega_n^2, \quad b_4 = 96\, c^2 \alpha^2 \omega_n^2, \quad b_5 = 0, \quad b_6 = -3\, \alpha^4$$

and

$$c_0 = 192\, c^3 \mu \omega_n^4, \quad c_1 = 64\, \omega_n^2 (\alpha^2 \mu^2 + 4\, c^4 \omega_n^2), \quad c_2 = 132\, c \alpha^2 \mu \omega_n^2,$$
$$c_3 = 96\, c^2 \alpha^2 \omega_n^2, \quad c_4 = 0, \quad c_5 = -3\, \alpha^4$$

Equation (4.8) is identical to Eq. (4.6), but now we also have an additional equation $\mathcal{G}_2(\sigma_{cr}, a_{cr}) = 0$. Once the value of a_{cr} is numerically computed, we substitute it into Eq. (4.9) to obtain the value of σ_{cr}. Like before, substituting the values of a_{cr} and σ_{cr} into either Eq. (4.4) or Eq. (4.5) gives us the critical forcing amplitude f_{cr}.

To determine the jump frequencies σ^*, we use the fact that $\mathcal{F}(a, \sigma) = 0$ and $\mathcal{F}'(a, \sigma) = 0$ at the jump points (σ^*, a^*). We begin again by computing a Gröbner basis for the polynomials $\mathcal{F}(a, \sigma)$ and $\mathcal{F}'(a, \sigma)$ and, thus, obtain two polynomials $\mathcal{G}_1$ and $\mathcal{G}_2$, which also vanish at the jump points (σ^*, a^*). Using either Eq. (4.4) or Eq. (4.5) and the lex order $\sigma > a$, we obtain

$$\mathcal{G}_1(a^*) = \sum_{i=0}^{12} b_i a^i \Big|_{a=a^*} = 0$$

$$\mathcal{G}_2(\sigma^*, a^*) = \sigma^* + \sum_{i=0}^{11} c_i a^i \Big|_{a=a^*} = 0$$

where the b_i and c_i are functions of known physical quantities. We solve for the values of a^* and σ^* numerically. But this time we do not obtain any spurious solutions of σ^* because of the unique form of $\mathcal{G}_2$. In this aspect, the Gröbner basis method can be viewed as a nonlinear version of the Gaussian elimination technique, which is used to solve linear polynomial equations.

4.2 Results

Following the procedure described in the previous section, we computed the critical forcing amplitude f_{cr} and the jump frequencies σ^* in the response of the cantilever beam for a value of $f > f_{cr}$. We used the **Resultant** and **Solve** functions of MATHEMATICA (Wolfram, 1999) to calculate the resultant of two polynomials and to compute roots of polynomials. For computing a Gröbner basis for two polynomials, we used the **GroebnerBasis** function. Identical solutions are obtained using the resultant and the Gröbner basis methods. The parameter values used in the calculations are: $\omega_n = 98\pi$, $\alpha = -7 \times 10^8$, $\zeta = 6 \times 10^{-4}$, $\int_0^l \Phi_n\, ds = 0.18$, and $c = 200$. The critical forcing amplitude is found to be $f_{cr} = 0.274$ with $\sigma_{cr} = -0.795$ ($\Omega_{cr} = 97.747\,\pi$) and $a_{cr} = 9.277 \times 10^{-4}$. Using Eq. (4.3), we obtain the frequency-response curve for $f = f_{cr}$, which is illustrated in Fig. 4.2(a). The asterisk in Fig. 4.2(a) denotes the inflection point (σ_{cr}, a_{cr}). For $a_b = 49$ ($f = 8.82$), the jump frequencies are found to be $\sigma^*_{up} = -9.199$ ($\Omega^*_{up} = 95.072\,\pi$) and $\sigma^*_{down} = -36.544$ ($\Omega^*_{down} = 86.368\,\pi$). The corresponding frequency-response curve is plotted, along with the computed jump-up and jump-down points in Fig. 4.2(b).

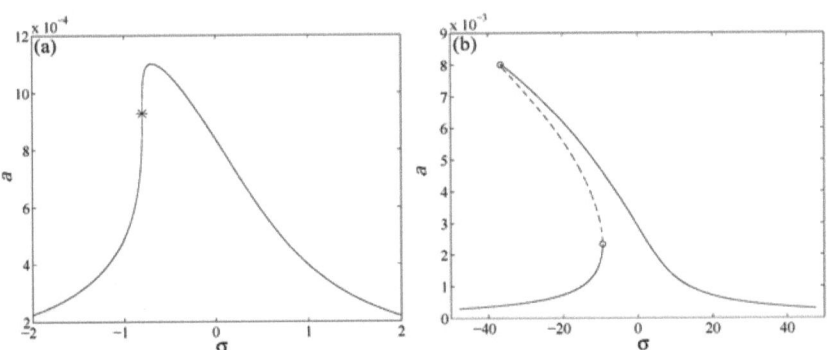

Figure 4.2: Frequency-response curves obtained using (a) $f = f_{cr}$ and (b) $f = 8.82$. The asterisk in (a) indicates the inflection point and the circles in (b) indicate the jump-up and jump-down points.

4.3 Closure

Knowing the form of the frequency-response function, one can easily and accurately determine the critical forcing amplitude and jump frequencies of a SDOF nonlinear system using the proposed methods. The only requirement being that the frequency-response function be a polynomial function in a and σ. The proposed simple and straightforward methods can be applied to a variety of systems. Also, the methods have the potential of being applicable to MDOF nonlinear systems.

Chapter 5

Energy Transfer Between Widely Spaced Modes Via Modulation

In this chapter, we study the transfer of energy between widely spaced modes (i.e., modes whose natural frequencies are wide apart) via modulation in flexible metallic cantilever beams. The presentation is divided into two parts – one dealing with planar motion and the other with nonplanar motion. In the first part, we present an experimental and theoretical study of the effect of excitation amplitude on nonresonant modal interactions. In particular, we study the response of a rectangular cross-section, flexible cantilever beam to a transverse excitation near its third natural frequency at various amplitudes of excitation. The transfer of energy between modes via modulation was also observed while directly exciting the fourth and fifth modes. But, in addition to a large first-mode response, a significant contribution from in-between modes was also observed. For simplicity, we therefore decided to excite the third mode, in which case we observed the presence of only the first mode in the event of an energy transfer. Also, the ratio of the first and third natural frequencies is close to 1:30; that is, the requirement for the transfer of energy between those modes via modulation is satisfied. Experimentally, we found transfer of energy from the high-frequency third mode to the low-frequency first mode, accompanied by a slow modulation of the amplitude and phase of the third mode. But with increasing amplitude of excitation, the transfer of energy to the first mode seemed to subside. A reduced-order analytical model is also developed to study the transfer of energy between the widely spaced modes. In the second part, we extend the planar reduced-order model to include out-of-plane modes and study the energy transfer between widely spaced modes in a circular rod under transverse excitation and in the presence of a one-to-one internal resonance. A comparison is also made between the experimental results obtained by S. Nayfeh and Nayfeh (1994) and the nonplanar reduced-order model results.

5.1 Planar Motion

5.1.1 Test Setup

A schematic of the experimental setup is shown in Fig. 5.1. The vertically mounted, slender, uniform cross-section, steel, cantilever beam has dimensions 662 mm × 12.71 mm × 0.55 mm. The density, shear modulus, and Young's modulus of the beam are taken as 7400 kg/m^3, 70 GPa, and 165.5 GPa, respectively. The beam is clamped to a 445 N shaker that provides an external (i.e., transverse to the axis of the beam) harmonic excitation at the base of the beam. The excitation is monitored by means of an accelerometer placed on the clamping fixture, and the response of the cantilever beam is measured using a 350 Ohm strain gage mounted approximately 35 mm from the fixed end of the beam. The strain gage is mounted near the root of the cantilever beam where the strains are high, and care has also been taken to keep it away from the strain nodes of the first five linear vibration modes. The strain gage and accelerometer signals are monitored, in both the frequency and time domains, by a digital signal analyzer, which is also used to drive the shaker. At points of interest, the time- and frequency-response data are stored onto a floppy disk for further characterization and processing. The frequency spectra of the strain gage and accelerometer signals are calculated in real time over a 12.5 Hz bandwidth (0.015 Hz frequency resolution) with a flat-top window. However, to measure the sideband spacing and the Hopf bifurcation frequency, we used a Hanning window. We waited for a long time to ensure steady state before taking any measurement.

In the case of a periodic response, the response amplitude of a mode can be determined from the frequency spectrum (or FFT), obtained using a flat-top window, by measuring the magnitude of the peak in the spectrum near the natural frequency of the mode. If the response of a mode is modulated, the frequency spectrum will contain sidebands and it would be difficult to accurately determine the amplitude from the spectrum. For simplicity, we continue to read the amplitude directly from the frequency spectrum even when the response is modulated.

Chapter 5. Energy Transfer

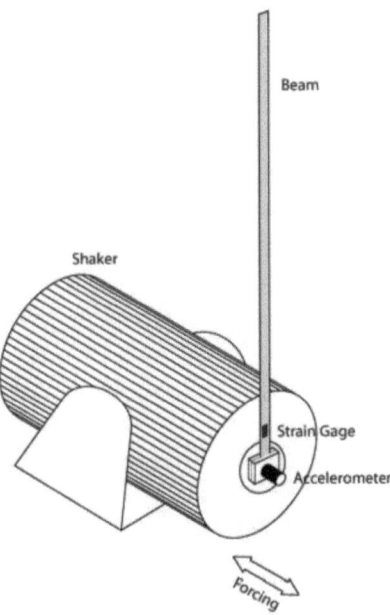

Figure 5.1: Experimental setup.

The strain gage basically measures the strain at the location where it is mounted on the beam. The output of the strain-gage conditioner is in volts, which can be easily converted into displacement for the case of a periodic, single-mode response. Owing to the complications due to modulations, as described in the above paragraph, all of the plots and FFTs are obtained using strain values only. However, there is no one-to-one correspondence between the strain values and the displacement amplitudes of the various modes; that is, one strain value could lead to different tip displacements for different modes. For instance, for a given strain value, the displacement of the beam tip, when the beam is vibrating only in the first mode, would be approximately eleven times compared to the case when the beam is vibrating only in the third mode.

The linear in-plane (i.e., in the plane of excitation) flexural natural frequencies of the beam were obtained using the frequency-response function of the signal analyzer. The beam was excited by a low-amplitude, 50% burst-chirp excitation, and a uniform window was used to analyze the power spectra of the accelerometer and strain-gage signals. Peaks in the amplitude portion of the frequency-response function give the linear natural frequencies. This process was repeated for several low-excitation levels until no noticeable shifts in the peaks were observed. The natural frequencies were also determined from an Euler-Bernoulli beam model incorporating the effect of gravity, which tends to lower the frequencies especially of the lower modes (Tabaddor and Nayfeh, 1997). The finite-element method (FEM) was used to solve the resulting model equation along with the corresponding boundary conditions. The experimentally and analytically obtained values of the first six linear in-plane flexural natural frequencies of the beam are listed in Table 5.1. It is clearly evident that the two sets of values match very closely. In addition, the first two out-of-plane (i.e., in the plane perpendicular to the plane of excitation) flexural natural frequencies of the beam were also determined using the FEM model and are found to be equal to 22.14 Hz and 138.84 Hz. The first torsional frequency of the slender beam is found to be equal to 100.52 Hz, assuming that the width of the beam is much greater than its thickness (Timoshenko and Goodier, 1970). The modal damping factors ζ_n were determined experimentally using the logarithmic decrement method. The first four damping factors are found to be equal to 9×10^{-3}, 1.85×10^{-3}, 2.25×10^{-3}, and 5×10^{-3}.

Table 5.1: The first six in-plane natural frequencies – experimental and analytical values.

Mode No.	Natural Frequency (Hz)	
	Experimental	Analytical
1	0.574	0.573
2	5.727	5.730
3	16.55	16.54
4	32.67	32.67
5	54.18	54.20
6	81.14	81.10

For completeness, we add that the beam is not perfectly straight, but has a small initial curvature, which could be due to the way it was manufactured and sold. Also, the shaker system has an inherent (low) quadratic nonlinearity, and there exists a shaker-beam interaction as is usually the case (McConnell, 1995). All of these factors could be affecting the beam response, but we assume that their influence on the response is negligible.

5.1.2 Experimental Results

Frequency- and force-response curves illustrate various characteristics of a nonlinear system like the presence of multiple stable responses, jumps, bifurcations, type of nonlinearity (softening or hardening), etc. So, as a first step, we obtained the frequency- and force-response curves for the third in-plane bending mode. For the frequency-response curve, the excitation amplitude a_b was held constant at $0.8g$, and the excitation frequency Ω was varied in the neighborhood of the third natural frequency. And for the force-response curve, the excitation frequency Ω was held constant at 16 Hz, and the excitation amplitude a_b was varied between 0 and $1g$. Changes in the control parameters (excitation frequency or amplitude) were made very gradually, and, at each value of the control parameter, transients were allowed to die out before the amplitude of the response was recorded. Data obtained from both forward and backward sweeps of the control parameter are used to plot the curves. In addition, to ensure that even isolated branches of the curves get located, we performed a third sweep where, at increments in the control parameter, we applied several disturbances to the beam in an effort to find all possible long-time responses. For certain frequency ranges, a small out-of-plane motion was also observed, which seemed to increase with an increase in the amplitude of the beam response. However, since it was small, we did not take any measurements of that motion.

The frequency-response curve of the third mode is shown in Fig. 5.2. Well away from the third natural frequency, the only mode present in the beam response is the third mode. This can be easily confirmed by a visual inspection of the beam motion. Also, the response spectrum shows only a single peak at the excitation frequency. As the frequency of excitation is swept downward from well above the third natural frequency, the third-mode response becomes modulated and a growing contribution of the low-frequency first mode is observed. This is the signature of energy transfer between widely spaced modes. Visually we can see the amplitude of the third mode being modulated, along with a large swaying (i.e., the first-mode response). Typical input and response time traces are illustrated in Fig. 5.3. We note here that strain (and not displacement) values are plotted and hence the first-mode response is much greater than what is observed in the response time trace in Fig. 5.3.

According to S. Nayfeh and Nayfeh (1993), the slow dynamics associated with the amplitude and phase of the high-frequency third mode interacts with the slow dynamics of the low-frequency first mode and eventually loses stability by a Hopf bifurcation, giving rise to amplitude and phase modulations of the third mode and creation of a new frequency close to the first-mode frequency. The amplitude and

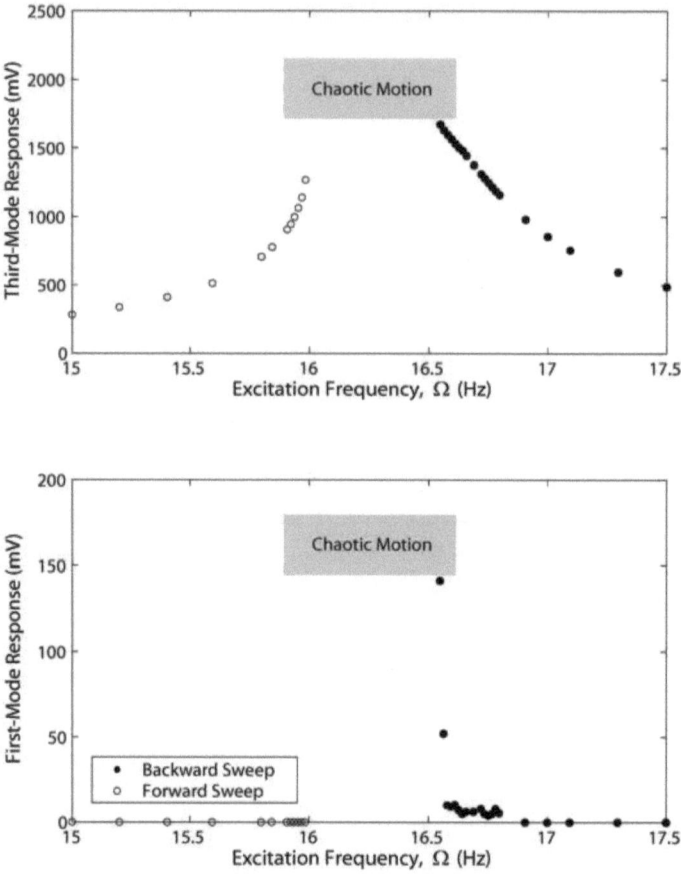

Figure 5.2: Frequency-response curve of the third mode when $a_b = 0.8g$.

phase modulations of the third mode are evident by the presence of asymmetric sidebands around the high-frequency component in the response spectrum. As the modulation is a result of an instability involving both the high- and low-frequency modes, the modulation frequency has to be equal to the newly created frequency due to the Hopf bifurcation. This newly created frequency will be henceforth referred to as the Hopf bifurcation frequency. By definition, the sideband spacing is equal to the

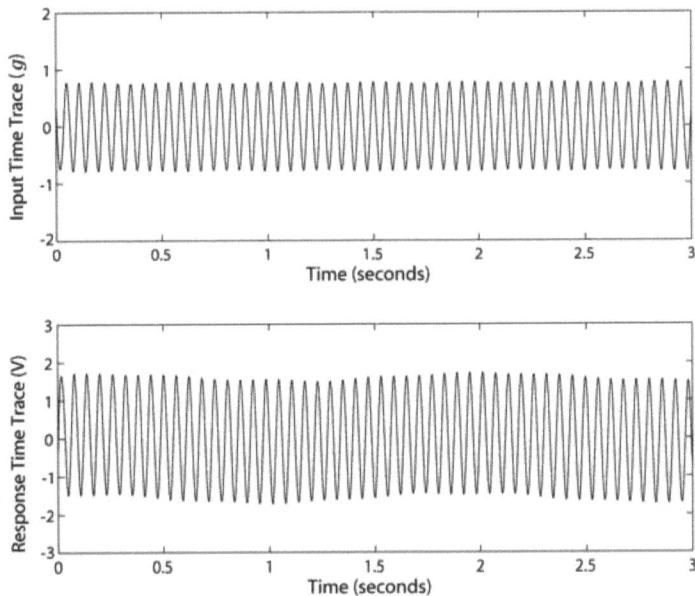

Figure 5.3: Input and response time traces at $\Omega = 16.547$ Hz when $a_b = 0.8g$.

modulation frequency.

The FFTs corresponding to the time traces in Fig. 5.3 are shown in Fig. 5.4. The response FFT shows two main peaks, one at the frequency of the excitation, which is near the third natural frequency, and the other at the Hopf bifurcation frequency. The Hopf bifurcation frequency is found to be equal to 0.547 Hz, which is close to the first-mode natural frequency. The asymmetric sideband structure around the peak corresponding to the third mode indicates that the response of the third mode is amplitude and phase modulated. Moreover, the sideband spacing (i.e., the modulation frequency) is equal to the Hopf bifurcation frequency. A close observation of the input spectrum reveals that the input is also modulated. This indicates a feedback from the structure to the shaker. As mentioned before, a structure-shaker interaction is more of a rule than an exception.

Decreasing the excitation frequency further, we observe the beam motion jump-up to a chaotically

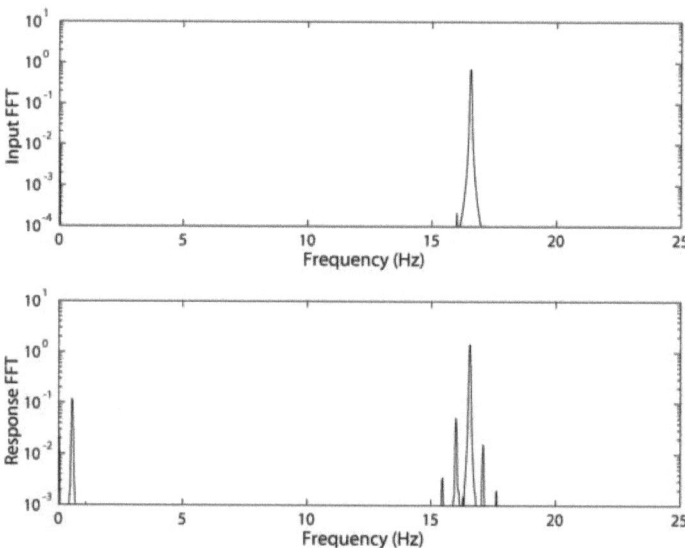

Figure 5.4: Input and response FFTs at $\Omega = 16.547$ Hz when $a_b = 0.8g$.

modulated motion. The modulation frequency and swaying amplitude increase with time and the beam response eventually gets drawn to a chaotic attractor. Figures 5.5(a) and 5.5(b) show the time trace of a transition to a chaotic motion and of a fully developed chaotic motion, respectively. The FFT of the fully developed chaotic motion is shown in Fig. 5.5(c). The FFT indicates a chaotic modulation of the responses of the third and first modes. In addition, we also see a chaotically modulated response of the second mode. Decreasing the excitation frequency even further, we observe the beam motion jump-down to a low-amplitude single-mode response consisting only of the third mode. In the forward sweep, starting from an excitation frequency well below the third natural frequency, we observe the beam motion jump-up from a low-amplitude third-mode response to a chaotic motion, and jump-down back to a low-amplitude third-mode response.

The force-response curve of the third mode is shown in Fig. 5.6. As the excitation amplitude is increased from zero, the beam motion eventually jumps from a periodic third-mode response to a chaotic motion directly. The FFT of such a chaotic motion indicates, like before, a chaotic modulation

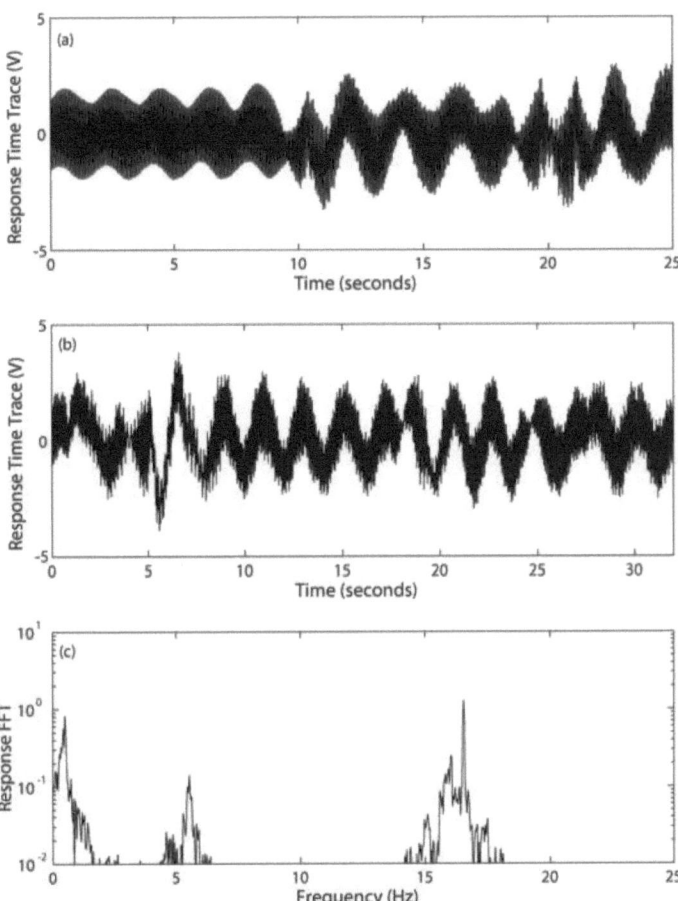

Figure 5.5: Time traces and FFT of the chaotic motion observed at $\Omega = 16.531$ Hz when $a_b = 0.8g$.

of the responses of the third, second, and first modes. In the backward sweep, the motion jumps from a chaotic response to a periodic third-mode response.

To study the influence of the excitation amplitude on the transfer of energy between widely spaced

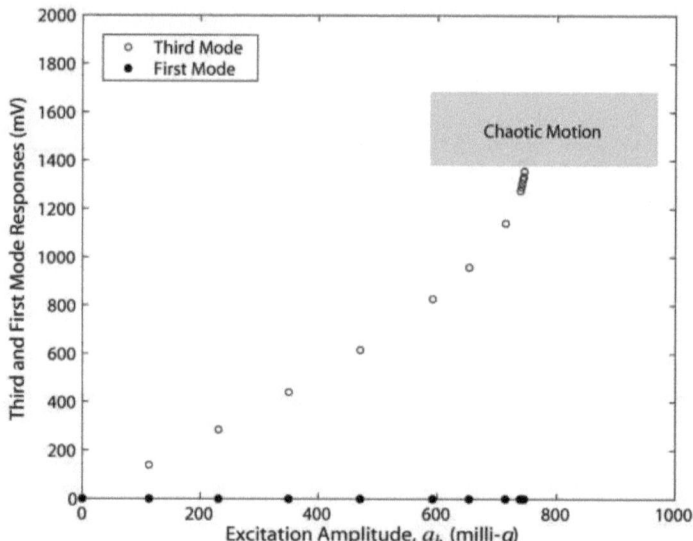

Figure 5.6: Force-response curve of the third mode when $\Omega = 16$ Hz.

modes, we repeated the above experiments at higher amplitudes of excitation. Figure 5.7 shows the input and response time traces at the excitation frequency $\Omega = 17.109$ Hz for the excitation amplitude $a_b = 2.3g$. The corresponding FFTs are shown in Fig. 5.8. We note that the Hopf bifurcation frequency and thus the sideband spacing has now increased to 1.375 Hz. Also, the amplitude of the peak at the Hopf bifurcation frequency is around ten times smaller compared to the case when the excitation amplitude a_b was equal to $0.8g$. The fact that the newly created Hopf bifurcation frequency is away from the natural frequency of the first mode seems to inhibit the transfer of energy from the third mode to the first mode. Consequently, we see very little swaying (i.e., the first-mode response), as is evident from the response time trace shown in Fig. 5.7. The sidebands around the peak of the excitation frequency in the input FFT, shown in Fig. 5.8, indicate modulation of the input.

Increasing the excitation amplitude a_b to $2.97g$, we observe a further increase in the Hopf bifurcation frequency to 1.578 Hz, resulting in an even lesser transfer of energy to the first mode. The input and response time traces and their corresponding FFTs at the excitation frequency $\Omega = 17.547$ Hz are

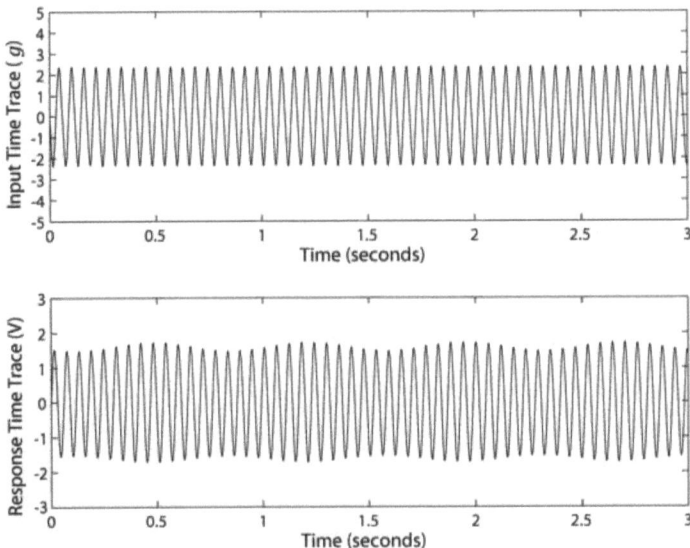

Figure 5.7: Input and response time traces at $\Omega = 17.109$ Hz when $a_b = 2.3g$.

shown in Figs. 5.9 and 5.10, respectively. For such a large excitation amplitude, the first-mode swaying is almost nil.

5.1.3 Reduced-Order Model

We develop a reduced-order analytical model to study the transfer of energy between widely spaced modes. In the analysis of a weakly damped, weakly nonlinear continuous system, which has an infinite number of degrees of freedom like the beam under study, a modal discretization is often employed to obtain a reduced-order model of the system (Nayfeh and Mook, 1979). The system response is expanded in terms of the undamped linear mode shapes multiplied by modal coordinates and substituted into the equation of motion. Then, the Galerkin's weighted residual method is employed to obtain a reduced-order model of the continuous system.

Modal discretization techniques essentially replace a set of partial-differential equations governing

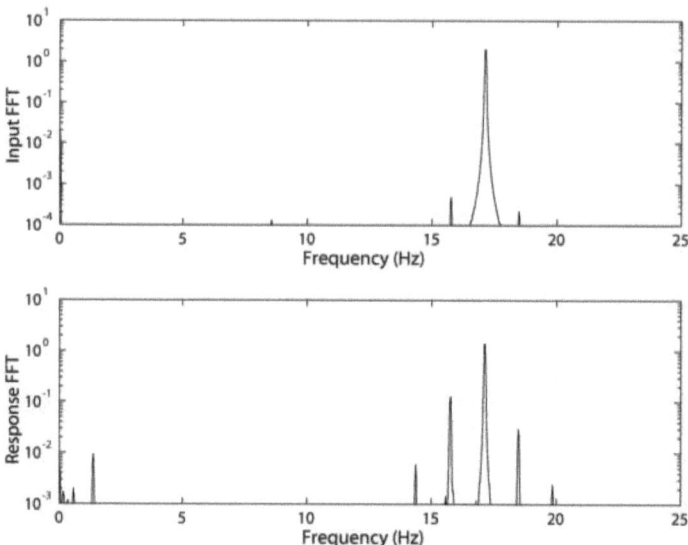

Figure 5.8: Input and response FFTs at $\Omega = 17.109$ Hz when $a_b = 2.3g$.

a continuous (infinite-dimensional) system with a finite set of nonlinearly coupled, ordinary-differential equations in terms of the modal coordinates. For simplicity and faster computation, a minimum number of modes necessary to represent the response are included in the expansion. However, one should ensure that the neglected modes do not affect the response of the system significantly, else the discretized system would lead to erroneous results.

Equations (2.58) and (2.59) governing the nonplanar dynamics of an isotropic, inextensional beam are simplified to the case of planar motion of a uniform metallic cantilever beam under external excitation. Thus, the governing equation reduces to

$$m\ddot{v} + c_v\dot{v} + EIv^{iv} = ma_b \cos\Omega t + mg[(s-l)v'' + v'] - EI\left[v'(v'v'')'\right]' - \frac{1}{2}m\left\{v'\int_l^s\left[\frac{\partial^2}{\partial t^2}\int_0^s v'^2 ds\right]ds\right\}'$$
(5.1)

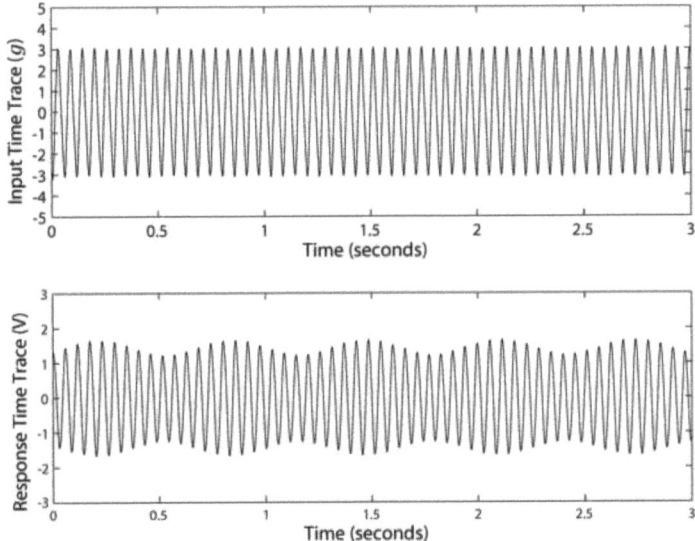

Figure 5.9: Input and response time traces at $\Omega = 17.547$ Hz when $a_b = 2.97g$.

and the associated boundary conditions are

$$v = 0 \quad \text{and} \quad v' = 0 \quad \text{at} \quad s = 0 \tag{5.2}$$

$$v'' = 0 \quad \text{and} \quad v''' = 0 \quad \text{at} \quad s = l \tag{5.3}$$

The overdot and prime indicate the derivatives with respect to time t and arclength s, respectively, $v(s, t)$ is the transverse displacement, m is the mass per unit length, l is the beam length, E is Young's modulus, I is the area moment of inertia, a_b is the acceleration of the supported end (base) of the beam, g ($= 9.8 \ m/s^2$) denotes the acceleration due to gravity, c_v is the coefficient of linear viscous damping per unit length, and Ω is the excitation frequency.

In Eq. (5.1), the first of the nonlinear terms on the right-hand side is a hardening nonlinearity arising from the potential energy stored in bending and is referred to as geometric nonlinearity. The second nonlinear term is a softening nonlinearity arising from the kinetic energy of axial motion and is usually referred to as inertia nonlinearity. For the third mode, the inertia nonlinearity is the dominant

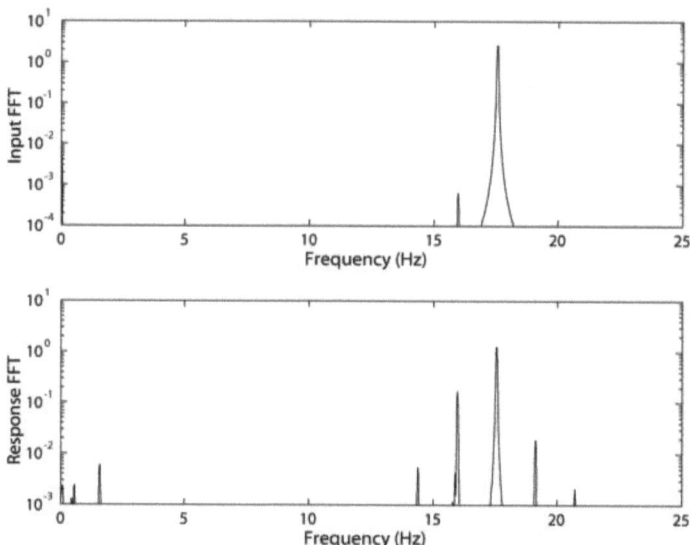

Figure 5.10: Input and response FFTs at $\Omega = 17.547$ Hz when $a_b = 2.97g$.

nonlinear term, whereas for the first mode, the geometric nonlinearity is the dominant nonlinear term.

The natural frequencies of the linear system corresponding to Eq. (5.1) are given by $\omega_n = r_n^2 \sqrt{EI/ml^4}$, where r_n is the nth root of the characteristic equation $1 + \cos(r)\cosh(r) = 0$. We introduce nondimensional variables, denoted by an asterisk, by using l as the characteristic length and the inverse of the third natural frequency ω_3 (= 16.824 Hz) as the characteristic time. Then, in nondimensional form, the governing equation and boundary conditions become

$$\ddot{v} + \mu \dot{v} + \frac{1}{r_3^4} v^{iv} = F\cos\Omega t + G[(s-1)v'' + v'] - \frac{1}{r_3^4}\left[v'(v'v'')'\right]' - \frac{1}{2}\left\{v'\int_1^s \left[\frac{\partial^2}{\partial t^2}\int_0^s v'^2 ds\right] ds\right\}' \quad (5.4)$$

$$v = 0 \quad \text{and} \quad v' = 0 \quad \text{at} \quad s = 0 \quad (5.5)$$

$$v'' = 0 \quad \text{and} \quad v''' = 0 \quad \text{at} \quad s = 1 \quad (5.6)$$

where the asterisks have been dropped for ease of notation, $F = ma_b/\gamma$ and $G = mg/\gamma$ with $\gamma = EIr_3^4/l^3$ and $\mu = l^2 c_v / r_3^2 \sqrt{mEI}$. With the chosen nondimensionalization, $\omega_3 = 1$ and $\omega_1 = r_1^2/r_3^2 = 0.057$. These frequency values indicate that the third and first modes are widely spaced.

As the beam constitutes a weakly nonlinear system, we expand its response $v(s,t)$ in terms of its undamped linear mode shapes as follows:

$$v(s,t) = \sum_{i=1}^{N} u_i(t)\phi_i(s) \tag{5.7}$$

where N denotes the number of retained modes, the $u_i(t)$ are generalized (modal) coordinates, and the $\phi_i(s)$ denote the normalized undamped linear mode shapes given by

$$\phi_i(s) = \cosh r_i s - \cos r_i s + \frac{\cos r_i + \cosh r_i}{\sin r_i + \sinh r_i}(\sin r_i s - \sinh r_i s) \tag{5.8}$$

Substituting Eq. (5.7) into Eq. (5.4), multiplying by ϕ_n, integrating the result over the length of the beam, and using the orthonormal properties of the linear mode shapes yields the set of equations

$$\ddot{u}_n + \mu_n \dot{u}_n + \omega_n^2 u_n = f_n \cos \Omega t + \sum_i g_{ni} u_i + \sum_{i,j,k} \Lambda_{nijk} u_i u_j u_k$$
$$+ \sum_{i,j,k} \Gamma_{nijk} u_k (\ddot{u}_i u_j + 2\dot{u}_i \dot{u}_j + u_i \ddot{u}_j), \quad n = 1, 2, \ldots, N \tag{5.9}$$

where

$$\Lambda_{nijk} = \frac{1}{r_3^4} \int_0^1 \phi'_n \phi'_i (\phi''_j \phi''_k + \phi'_j \phi'''_k) ds$$

and

$$\Gamma_{nijk} = -\frac{1}{2} \int_0^1 \left(\int_0^s \phi'_n \phi'_k ds \right) \left(\int_0^s \phi'_i \phi'_j ds \right) ds$$

are the coefficients of the cubic geometric and inertia nonlinearity terms, respectively, in the discretized equations, and

$$\mu_n = \int_0^1 \mu \phi_n^2 ds, \quad f_n = \int_0^1 F\phi_n ds, \quad g_{ni} = \int_0^1 G\phi_n [(s-1)\phi''_i + \phi'_i]ds, \quad \sum \equiv \sum_{i,j,k} \sum_i \sum_j \sum_k$$

Using $c_v = 2m\zeta_n \omega_n^{ex}$, where ζ_n is the damping factor of the nth mode and ω_n^{ex} is the experimentally determined nth natural frequency, we obtain

$$\mu_n = 2\zeta_n \frac{r_n^2}{r_3^2} \frac{\omega_n^{ex}}{\omega_n}$$

In the experiments, we observed, during the energy transfer, that the response spectrum consists essentially of peaks near the first and third natural frequencies (refer to Fig. 5.4), with sidebands around the latter. It is thus natural to include only the third and first modes in the expansion in Eq. (5.7). But we found out that not including the second and fourth modes led to results inconsistent with the experimental observations. We, therefore, retain the first four modes in the expansion of $v(s,t)$; that is, $N = 4$. Some nonlinear terms in the discretized equations contain $\ddot{u}_n$ in addition to u_n and $\dot{u}_n$, as seen in Eq. (5.9). Solving the four discretized equations for $\ddot{u}_n$ ($n = 1, 2, 3, 4$) in terms of u_n, $\dot{u}_n$, and t, and using the state-space approach, we obtain the set of eight first-order ordinary-differential equations

$$\dot{\mathbf{x}} = F(\mathbf{x}, t) \qquad (5.10)$$

where

$$\mathbf{x} = \{x_1, x_2, x_3, x_4, x_5, x_6, x_7, x_8\}^T = \{u_1, \dot{u}_1, u_2, \dot{u}_2, u_3, \dot{u}_3, u_4, \dot{u}_4\}^T$$

and

$$F(\mathbf{x}, t) = \{x_2, \ddot{u}_1(\mathbf{x}, t), x_4, \ddot{u}_2(\mathbf{x}, t), x_6, \ddot{u}_3(\mathbf{x}, t), x_8, \ddot{u}_4(\mathbf{x}, t)\}^T$$

The set of first-order ODEs can be easily solved using the following algorithms: Runge-Kutta-Fehlberg or Adams for non-stiff systems and Gear for stiff systems. But the system of equations is stiff, as the first mode evolves on a slow scale while the third mode evolves on a fast scale. Thus, the Gear algorithm would be ideal for solving the system of equations given by Eq. (5.10). Once we know the steady-state values of the u_i, we can substitute them into Eq. (5.7) and determine the transverse displacement v of the beam.

5.1.4 Numerical Results

We now present results obtained by integrating the set of Eq. (5.10). Figure 5.11 shows the time trace of the displacement of the beam, at the location where the strain gage is mounted, and the corresponding FFT at the excitation frequency $\Omega = 0.977$ ($= 16.429$ Hz) for the excitation amplitude $a_b = 1.5g$. From the FFT and the time trace, it is obvious that the third-mode response is modulated and that there is a first-mode component in the response. The modulation frequency (i.e., the sideband spacing) is equal to the newly created Hopf bifurcation frequency, which is equal to 0.592 Hz. This is, therefore, similar to the energy transfer observed earlier in the experiments. Also, it is evident from the time trace that there is a static component present in the displacement v. This is similar to the results obtained by

S. Nayfeh and Nayfeh (1993) using their representative model. Keeping the excitation amplitude a_b fixed and lowering the excitation frequency Ω to 0.945 leads to a chaotically modulated motion. The associated time trace of the displacement and its FFT are shown in Fig. 5.12. Akin to the experimental results, we see, in the FFT, a peak near the second-mode frequency.

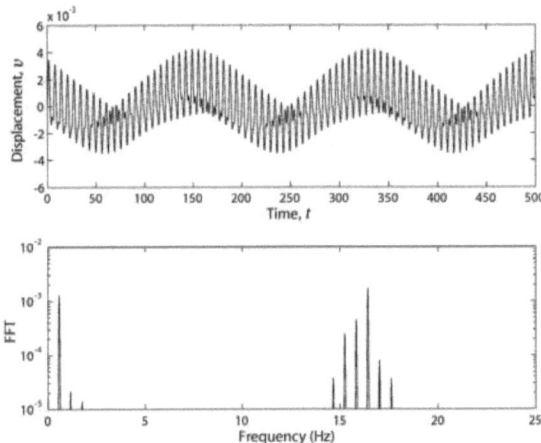

Figure 5.11: Displacement time trace and FFT at $\Omega = 0.977$ when $a_b = 1.5g$.

To study the effect of the excitation amplitude a_b on the modulation frequency and thus the energy transfer from the third mode to the first mode, we integrated Eq. (5.10) for three different values of a_b, namely, $1g$, $2g$, and $2.5g$. The corresponding Hopf bifurcation frequencies are found to be 0.571, 0.609, and 0.627 Hz, respectively. The response FFTs are shown in Fig. 5.13. Thus, with an increase in the excitation amplitude, the analytical model predicts a nominal increase in the Hopf bifurcation frequency. Whereas in the experiments, the increase in the Hopf bifurcation frequency with an increase in the excitation amplitude is substantial. Also, the model predicts that the amplitude of the first-mode component increases with an increase in the excitation amplitude. This is in contradiction with the trend observed in the experiments.

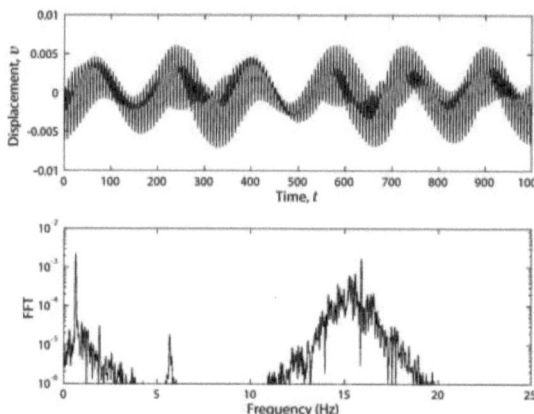

Figure 5.12: Displacement time trace and FFT at $\Omega = 0.945$ when $a_b = 1.5g$.

5.2 Nonplanar Motion

5.2.1 Experiments with a Circular Rod

In experiments with a slender, circular cross-section, steel, cantilever beam, S. Nayfeh and Nayfeh (1994) observed the transfer of energy between widely spaced modes via modulation when the rod was transversely excited near the natural frequency of its fifth (in fact third or any higher) mode. Here we briefly talk about those experiments and the results obtained by S. Nayfeh and Nayfeh. The length and diameter of the cantilever, used in the experiments, are 34.5 in and 0.0625 in, respectively, and its first five linear natural frequencies are 1.303, 9.049, 25.564, 50.213, and 83.150 Hz. The ratio of the natural frequencies of the first and fifth in-plane (or out-of-plane) modes is around 1:64. Because of axial symmetry, one-to-one internal resonances occur at each natural frequency of the beam, and the mode in the plane of excitation interacts with the out-of-plane mode of equal frequency, resulting in nonplanar whirling motions. For a certain range of parameters, a large-amplitude first-mode response was accompanied by a modulation of the amplitude and phase of the fifth-mode response, with the modulation frequency being approximately equal to the natural frequency of the first mode.

The frequency-response curves for the fifth in-plane and out-of-plane modes of the circular rod are

Chapter 5. Energy Transfer

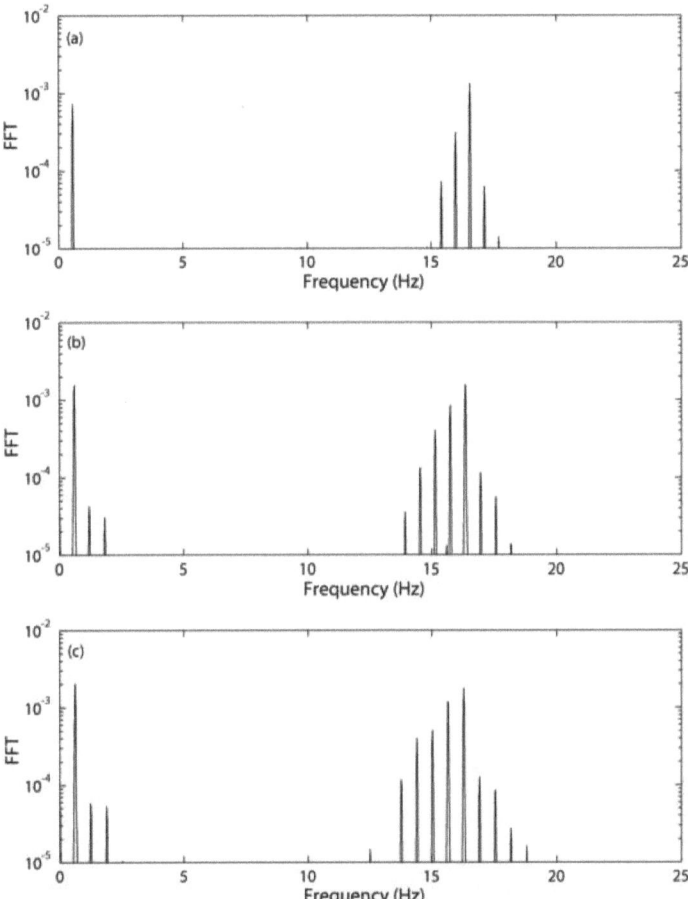

Figure 5.13: Displacement FFTs at (a) $\Omega = 0.984$ when $a_b = 1g$, (b) $\Omega = 0.972$ when $a_b = 2g$, and (c) $\Omega = 0.9678$ when $a_b = 2.5g$.

shown in Fig. 5.14. The excitation level was held constant at $2g$ rms and the excitation frequency was varied in the neighborhood of the fifth natural frequency. The data in the plots is a composite of the responses obtained by performing both backward and forward frequency sweeps. In the following paragraphs, we discuss in more detail both the weakly- and strongly-modulated motions observed during the sweeps.

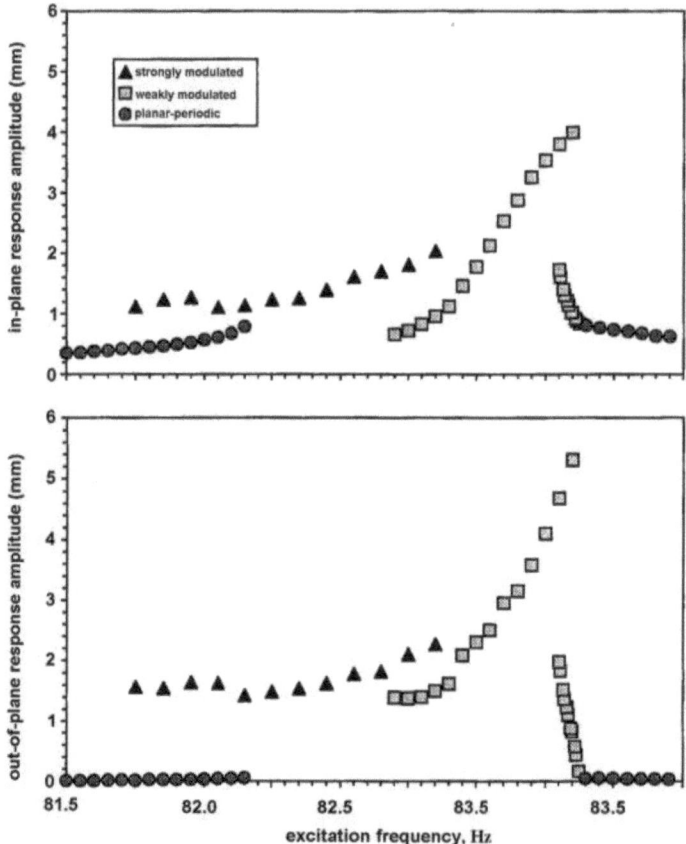

Figure 5.14: Frequency-response curves of the fifth mode of a circular rod for an excitation amplitude of $2g$ rms (S. Nayfeh and Nayfeh, 1994).

The observed weakly-modulated responses contained a large low-frequency component superimposed on a nearly constant amplitude fifth-mode whirling motion. Typical short and long time traces

Chapter 5. Energy Transfer   89

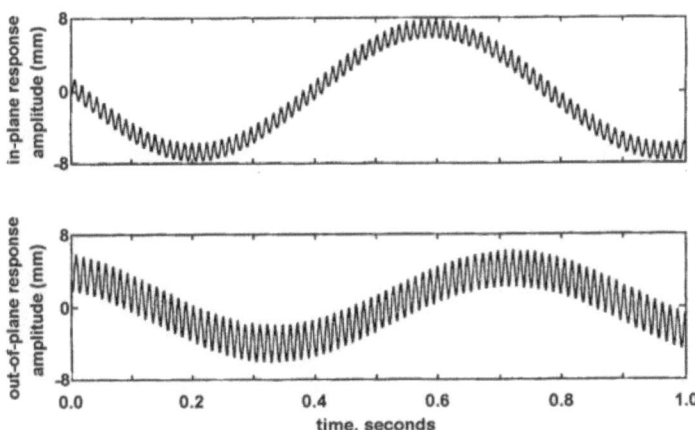

Figure 5.15: A short portion of the long time history of a typically weakly modulated motion of a circular rod (S. Nayfeh and Nayfeh, 1994).

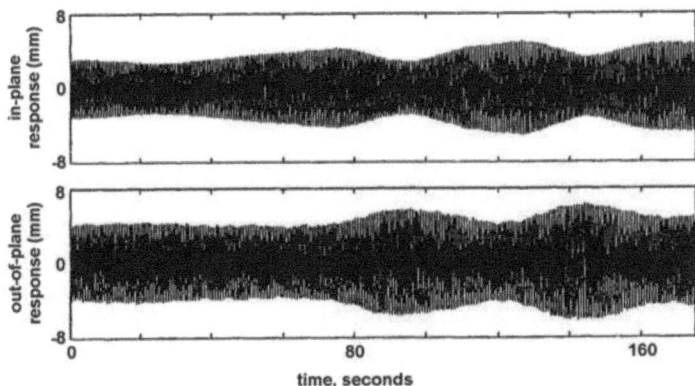

Figure 5.16: Time traces of a typically weakly modulated motion of a circular rod (S. Nayfeh and Nayfeh, 1994).

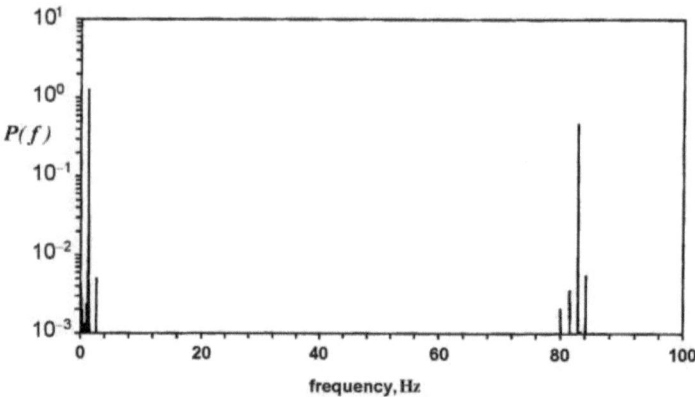

Figure 5.17: Power spectrum of a typically weakly modulated motion of a circular rod (S. Nayfeh and Nayfeh, 1994).

of in-plane and out-of-plane responses of this type are shown in Figs. 5.15 and 5.16. The long time traces illustrate the extremely slow variation of the amplitude of the first mode, but the short time traces do not readily reveal any modulation of the high-frequency component of the response. A typical FFT of this type of response is shown in Fig. 5.17. The FFT shows two main peaks, one at the frequency of excitation (near the fifth natural frequency) and the other close to the natural frequency of the first mode. Sidebands around the peak corresponding to the fifth mode indicate that the response of the fifth mode is modulated. Moreover, the sideband spacing is close to the first natural frequency. As indicated by the dense set of sidebands clustered around the peak near the first natural frequency, the response of the first mode is also modulated.

The most obvious feature of the strongly-modulated motions is the modulation of the amplitude of the fifth mode. Typical short and long time traces of this type of motion are shown in Figs. 5.18 and 5.19. In constrast to the case of the weakly-modulated motions of Fig. 5.15, the modulation of the fifth mode is clearly distinguishable in Fig. 5.18. The erratic behavior of the rod response in Fig. 5.19 suggests that the fifth mode is chaotically modulated, which can be clearly seen by the narrow band of response present in the neighborhood of the fifth natural frequency in the corresponding FFT shown in Fig. 5.20. As in the case of the weakly-modulated motions, there appears a dense set of sidebands clustered about the first-natural frequency peak, indicating that the first-mode response is also modulated.

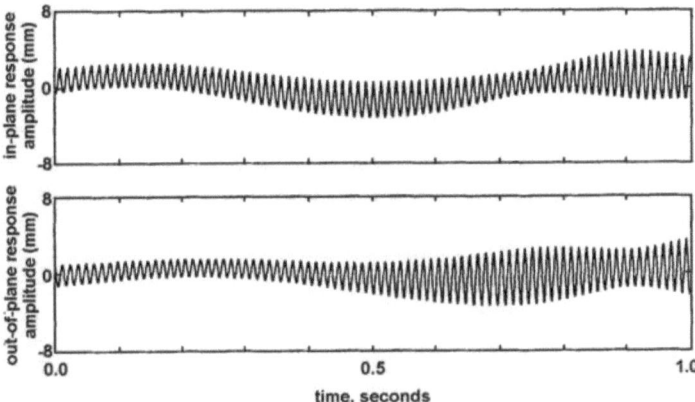

Figure 5.18: A short portion of the long time history of a typically strongly modulated motion of a circular rod (S. Nayfeh and Nayfeh, 1994).

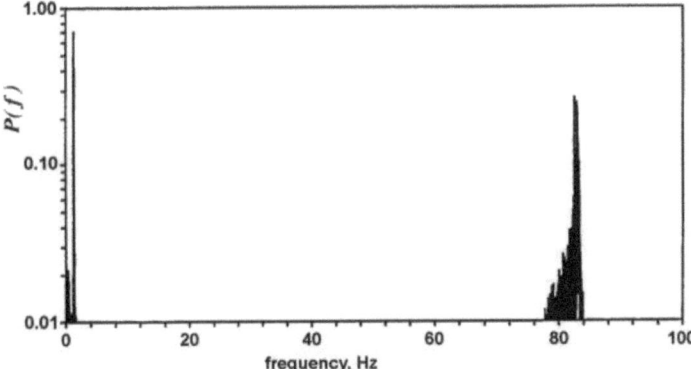

Figure 5.19: Time traces of a typically strongly modulated motion of a circular rod (S. Nayfeh and Nayfeh, 1994).

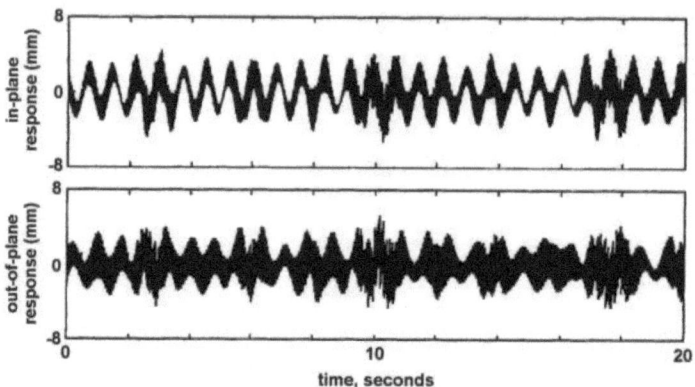

Figure 5.20: Power spectrum of a typically strongly modulated motion of a circular rod (S. Nayfeh and Nayfeh, 1994).

5.2.2 Reduced-Order Model

Now we develop a reduced-order analytical model to study the transfer of energy between widely spaced modes via modulation in the presence of a one-to-one internal resonance. This time the out-of-plane beam motion will also be included in the model. Equations (2.58) and (2.59) governing the nonplanar, nonlinear dynamics of isotropic, inextensible, Euler-Bernoulli beams will be used to study the response of the uniform, circular cross-section, metallic cantilever rod used by S. Nayfeh and Nayfeh (1994) in their experiments. We introduce nondimensional variables, denoted by an asterisk, by using l as the characteristic length and the inverse of the fifth natural frequency ω_5 (= 83.086 Hz) as the characteristic time. Then, in nondimensional form, the governing equations and boundary conditions become

$$\ddot{v} + \mu_v \dot{v} + \frac{1}{r_5^4} v^{iv} = F\cos\Omega t + G[(s-1)v'' + v'] - \frac{1}{r_5^4}\left[v'(v'v'')' + v'(w'w'')'\right]'$$
$$- \frac{1}{2}\left\{v' \int_1^s \left[\frac{\partial^2}{\partial t^2} \int_0^s (v'^2 + w'^2)ds\right]ds\right\}' \quad (5.11)$$

$$\ddot{w} + \mu_w \dot{w} + \frac{1}{r_5^4} w^{iv} = G[(s-1)w'' + w'] - \frac{1}{r_5^4}\left[w'(w'w'')' + w'(v'v'')'\right]'$$
$$- \frac{1}{2}\left\{w' \int_1^s \left[\frac{\partial^2}{\partial t^2} \int_0^s (v'^2 + w'^2)ds\right]ds\right\}' \quad (5.12)$$

$$v = 0, \quad v' = 0, \quad w = 0, \quad \text{and} \quad w' = 0 \quad \text{at} \quad s = 0 \tag{5.13}$$

$$v'' = 0, \quad v''' = 0, \quad w'' = 0, \quad \text{and} \quad w''' = 0 \quad \text{at} \quad s = 1 \tag{5.14}$$

where the asterisks have been dropped for ease of notation, $F = ma_b/\gamma$ and $G = mg/\gamma$ with $\gamma = EIr_5^4/l^3$, $\mu_v = l^2 c_v/r_5^2 \sqrt{mEI}$ and $\mu_w = l^2 c_w/r_5^2 \sqrt{mEI}$. With the chosen nondimensionalization, $\omega_5 = 1$ and $\omega_1 = r_1^2/r_5^2 = 0.0176$. These frequency values indicate that the fifth and first modes are widely spaced.

As the beam constitutes a weakly nonlinear system, we expand its in-plane and out-plane responses, $v(s,t)$ and $w(s,t)$, in terms of its undamped linear mode shapes as follows:

$$v(s,t) = \sum_i v_i(t)\phi_i(s) \tag{5.15}$$

$$w(s,t) = \sum_i w_i(t)\phi_i(s) \tag{5.16}$$

where the $v_i(t)$ and $w_i(t)$, $i = 1, 5$, are generalized (modal) coordinates and the $\phi_i(s)$ denote the normalized undamped linear mode shapes given by Eq. (5.8). We have included only two modes, in each expansion, to keep the model simple.

Substituting Eqs. (5.15) and (5.16) into Eqs. (5.11) and (5.12), multiplying by ϕ_n, integrating the result over the length of the beam, and using the orthonormal properties of the linear mode shapes yields the set of equations

$$\ddot{v}_n + \mu_{vn}\dot{v}_n + \omega_n^2 v_n = f_n \cos\Omega t + \sum_i g_{ni}v_i + \sum_{i,j,k} \Lambda_{nijk}(v_i v_j v_k + v_i w_j w_k)$$
$$+ \sum_{i,j,k} \Gamma_{nijk} v_k(\ddot{v}_i v_j + 2\dot{v}_i \dot{v}_j + v_i \ddot{v}_j + \ddot{w}_i w_j + 2\dot{w}_i \dot{w}_j + w_i \ddot{w}_j), \quad n = 1, 5 \tag{5.17}$$

$$\ddot{w}_n + \mu_{wn}\dot{w}_n + \omega_n^2 w_n = \sum_i g_{ni}w_i + \sum_{i,j,k} \Lambda_{nijk}(w_i v_j v_k + w_i w_j w_k)$$
$$+ \sum_{i,j,k} \Gamma_{nijk} w_k(\ddot{v}_i v_j + 2\dot{v}_i \dot{v}_j + v_i \ddot{v}_j + \ddot{w}_i w_j + 2\dot{w}_i \dot{w}_j + w_i \ddot{w}_j), \quad n = 1, 5 \tag{5.18}$$

where

$$\Lambda_{nijk} = \frac{1}{r_5^4} \int_0^1 \phi_n' \phi_i' (\phi_j'' \phi_k'' + \phi_j' \phi_k''') ds$$

and

$$\Gamma_{nijk} = -\frac{1}{2} \int_0^1 \left(\int_0^s \phi_n' \phi_k' ds \right) \left(\int_0^s \phi_i' \phi_j' ds \right) ds$$

are the coefficients of the cubic geometric and inertia nonlinearity terms, respectively, in the discretized equations, and

$$\mu_{vn} = \int_0^1 \mu_v \phi_n^2 ds, \quad \mu_{wn} = \int_0^1 \mu_w \phi_n^2 ds, \quad f_n = \int_0^1 F \phi_n ds, \quad g_{ni} = \int_0^1 G \phi_n [(s-1)\phi_i'' + \phi_i'] ds$$

Using $c_v = 2m\zeta_{vn}\omega_n^{ex}$ and $c_w = 2m\zeta_{wn}\omega_n^{ex}$, where ζ_{vn} and ζ_{wn} are the damping factors of the nth in-plane and out-of-plane modes, respectively, and ω_n^{ex} is the experimentally determined nth natural frequency, we obtain

$$\mu_{vn} = 2\zeta_{vn} \frac{r_n^2}{r_5^2} \frac{\omega_n^{ex}}{\omega_n} \quad \text{and} \quad \mu_{wn} = 2\zeta_{wn} \frac{r_n^2}{r_5^2} \frac{\omega_n^{ex}}{\omega_n}$$

Some nonlinear terms in the discretized equations contain $\ddot{v}_n$ and $\ddot{w}_n$ in addition to v_n, w_n, $\dot{v}_n$, and $\dot{w}_n$. Solving the four discretized equations for $\ddot{v}_n$ and $\ddot{w}_n$ ($n = 1, 5$) in terms of v_n, w_n, $\dot{v}_n$, $\dot{w}_n$, and t, and using the state-space approach, we obtain the set of eight first-order ordinary-differential equations

$$\dot{\mathbf{x}} = F(\mathbf{x}, t) \tag{5.19}$$

where

$$\mathbf{x} = \{x_1, x_2, x_3, x_4, x_5, x_6, x_7, x_8\}^T = \{v_1, \dot{v}_1, v_5, \dot{v}_5, w_1, \dot{w}_1, w_5, \dot{w}_5\}^T$$

and

$$F(\mathbf{x}, t) = \{x_2, \ddot{v}_1(\mathbf{x}, t), x_4, \ddot{v}_5(\mathbf{x}, t), x_6, \ddot{w}_1(\mathbf{x}, t), x_8, \ddot{w}_5(\mathbf{x}, t)\}^T$$

We integrate Eq. (5.19) for a long time to determine the steady-state values of the v_i and w_i ($i = 1, 5$). Once we know the values of the v_i and w_i, we can substitute them into Eqs. (5.15) and (5.16) and determine the in-plane and out-of-plane transverse displacements of the beam; that is, v and w.

5.2.3 Numerical Results

We now present results obtained by integrating the set of Eq. (5.19). We keep the excitation amplitude a_b constant at $2g$ rms and vary the excitation frequency Ω around the fifth natural frequency ω_5. The values used for the various parameters appearing in the equations are as follows: damping factors $\zeta_{v1} = \zeta_{w1} = 7.5 \; 10^{-5}$ and $\zeta_{v5} = \zeta_{w5} = 2.5 \; 10^{-3}$; elasticity modulus $E = 200$ GPa; and density

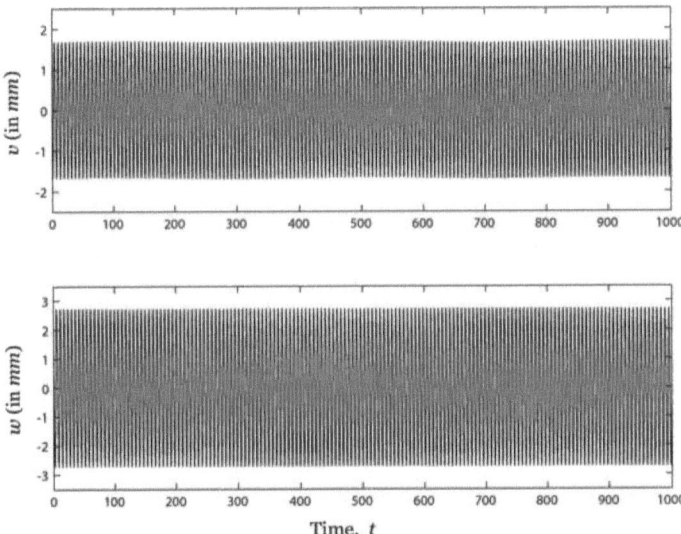

Figure 5.21: Time traces of in-plane and out-of-plane motion of beam at $\Omega = 82.75$ Hz.

$\rho = 7830\ kg/m^3$. Figure 5.21 shows the time traces of the in-plane and out-of-plane displacements of the beam tip at the excitation frequency $\Omega = 82.75$ Hz and the corresponding FFTs are shown in Fig. 5.22. From the FFTs shown in Fig. 5.22, it is clear that the fifth-mode response is modulated and that there is a first-mode component in the response. The modulation frequency (i.e., the sideband spacing) is equal to 1.3 Hz, which is close to the first natural frequency ($= 1.303$ Hz). But the amplitude of the first-mode component is very small in comparison to that of the fifth mode, which is apparent from the time traces and FFTs. This motion is qualitatively similar to the weakly modulated motion found experimentally by S. Nayfeh and Nayfeh (1994).

Reducing the excitation frequency resulted in a chaotically modulated motion. The time traces and FFTs of such a motion at $\Omega = 82.59$ Hz are shown in Figs. 5.23 and 5.24, respectively. From the FFTs, it is clear that both the first and fifth-mode responses are chaotically modulated. This motion is qualitatively similar to the strongly modulated motion found experimentally by S. Nayfeh and Nayfeh (1994).

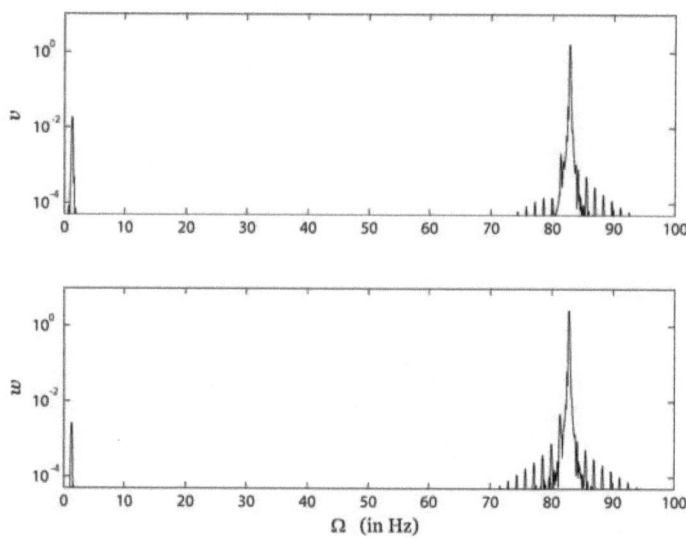

Figure 5.22: FFTs of in-plane and out-of-plane motion of beam at $\Omega = 82.75$ Hz.

5.3 Closure

We observed experimentally that the transfer of energy between widely spaced modes is a function of the closeness of the modulation frequency to the natural frequency of the first mode. The modulation frequency, which depends on various parameters like the amplitude and frequency of excitation, damping factors, etc., has to be near the natural frequency of the first mode for significant transfer of energy from the directly excited high-frequency third mode to the low-frequency first mode. In such a case, visually we see a large swaying of the beam due to the large amplitude of the first-mode component in the beam response. This is akin to primary resonance in a structural system where the closeness of an external excitation frequency to one of its natural frequencies dictates the magnitude of the structure response. When the excitation frequency is close to a natural frequency of the structure, the structure vibrates with a large amplitude, otherwise the amplitude is small.

Using the planar reduced-order model for the rectangular beam, we demonstrated the transfer of energy from the high-frequency third mode to the low-frequency first mode. Also, we found out that the

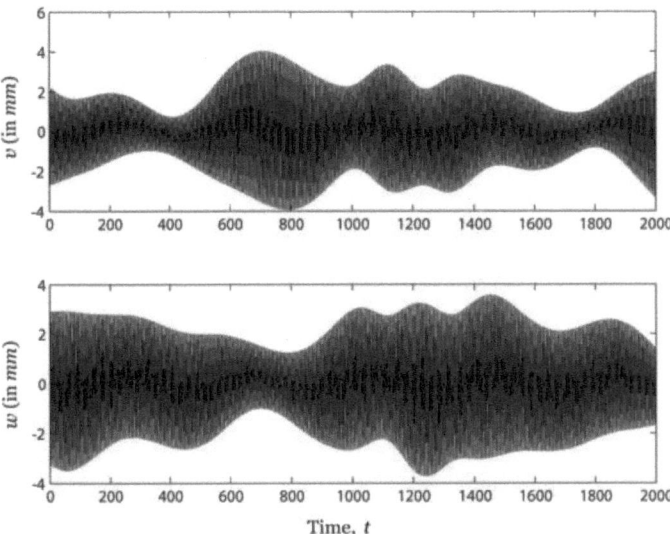

Figure 5.23: Time traces of in-plane and out-of-plane motion of beam at $\Omega = 82.59$ Hz.

first four modes need to be included in the Galerkin discretization for accurate modeling of the beam system. This shows that care should be taken while deciding on the number of modes to be retained in the approximation. The discrepancy in the experimental and analytical results can be attributed to a couple of shortcomings in the analytical model. Firstly, the out-of-plane motion is not accounted for in the model. In the experiments, a small out-of-plane motion is observed. Also, the natural frequency of the first out-of-plane mode is 22.14 Hz, which is somewhat close to the values used for the excitation frequency Ω. Therefore, including the out-of-plane motion in the analytical model might lead to better results. Secondly, the initial curvature in the beam might produce quadratic nonlinearities, which need to be included in the analytical model.

Using the nonplanar reduced-order model for the circular rod, we demonstrated the transfer of energy from the high-frequency fifth mode to the low-frequency first-mode in the presence of a one-to-one internal resonance. The model was able to predict results qualitatively similar to the weakly and strongly modulated motions observed experimentally by S. Nayfeh and Nayfeh (1994). But the

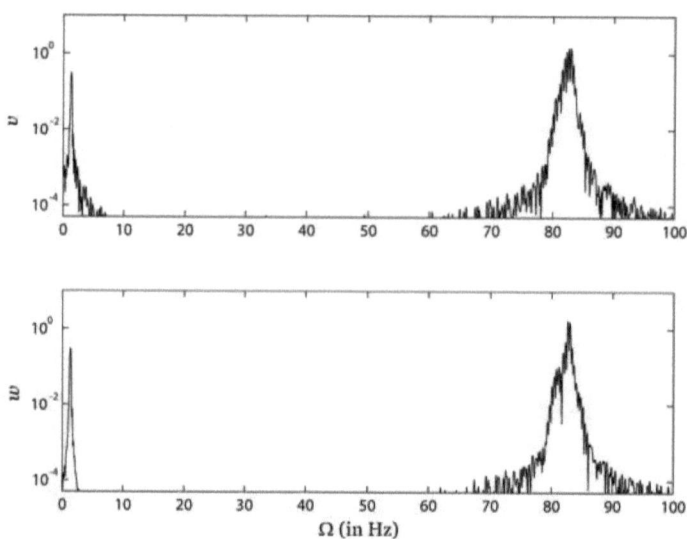

Figure 5.24: FFTs of in-plane and out-of-plane motion of beam at $\Omega = 82.59$ Hz.

reduced-order model predicted, for the amplitude of the first-mode component, values much smaller than those observed experimentally. To increase the amplitude of the first-mode component, we tried varying the damping factors, but in vain. This points to some inadequacies in the model. Issues, like the shaker-structure interaction, nonlinear damping, initial curvature of the beam, extension of the beam axis, and inclusion of more modes in the modal discretization, were ignored while developing the model. It is possible that one or more of these factors might be playing a bigger role than expected.

In addition, we also tried to obtain approximate analytical solutions for the beam response using two perturbation techniques, namely the methods of multiple scales and averaging. Using these two methods, we could not even demonstrate the transfer of energy from a high-frequency to a low-frequency mode via modulation. Further investigation needs to be carried out to determine the cause of such a discrepancy.

Chapter 6

Experiments with a Cantilever Plate

Like beams, rectangular plates are also one of the most commonly used structures in engineering applications. Many experimental studies have been done so far on rectangular plates, both metallic and composite, and a variety of nonlinear phenomena observed (Yamaki and Chiba, 1983; Yamaki, Otomo, and Chiba, 1983; Cole, 1990; Ostiguy and Evan-Iwanowski, 1993; Oh, 1994; Oh and Nayfeh, 1998). For a detailed review of the nonlinear vibration of plates, we refer the reader to Nayfeh and Mook (1979), Chia (1980), Sathyamoorthy (1997), and Nayfeh (2000).

In this chapter, an experimental study of the response of a thin, rectangular, aluminum cantilever plate under transverse harmonic excitation is presented. Three test sequences, each involving a frequency sweep around a particular natural frequency of the plate and a different amplitude of base excitation, are performed. A couple of interesting nonlinear dynamics phenomena present themselves in every test sequence. These phenomena include two-to-one and three-to-one internal resonances, external combination resonance, energy transfer between widely spaced modes, period-doubled motions, and chaos. In addition, the influence of the excitation amplitude on the energy transfer between widely spaced modes via modulation is also investigated.

6.1 Test Setup

A schematic of the experimental setup is shown in Fig. 6.1. The uniform cross-section, cantilever aluminum plate has dimensions $9'' \times 7.5'' \times 0.03''$ ($22.86\ cm \times 19.05\ cm \times 0.08\ cm$). The plate is clamped in a horizontal position to a 2000-lb shaker that provides an external (i.e., transverse to the plane of the plate) harmonic excitation at the fixed edge of the plate. The plate is not perfectly straight, but has some initial curvature. The difference in height of the clamped edge and the opposite free edge of the plate is around $0.4''$ ($1\ cm$). The shaker excitation is monitored by means of an accelerometer placed on the clamping fixture, and the response of the cantilever plate is measured using two strain gages mounted approximately $0.25''$ ($0.64\ cm$) from the fixed edge and at distances $2.25''$ ($5.72\ cm$) and $3.75''$ ($9.53\ cm$) from the lower left corner of the plate. If a strain gage is located along a nodal line of a mode, then it would not detect that mode at all. To avoid such a situation, we used two strain gages and compared, during every reading, their data to ensure that the contributions of all of the participating modes were being accounted. The signals of the accelerometer and strain gages are monitored, in both the frequency and time domains, by a digital signal analyzer, which is also used to drive the shaker. At points of interest, the time- and frequency-response data are stored onto a floppy disk for further characterization and processing. We waited for a long time to ensure steady state before taking any measurement. The strain gages basically measure the strain at the location

Figure 6.1: Experimental setup.

where they are mounted on the plate. However, it is difficult to accurately determine the displacement amplitudes of the various modes from the outputs of the strain-gage conditioners, which are in volts, especially when there is a modal interaction. Therefore, all of the plots and FFTs are obtained using strain values only. However, we note that there is no one-to-one correspondence between the strain values and the displacement amplitudes of the various modes; that is, one strain value could lead to different displacements for different modes.

A rough estimate of the natural frequencies of the plate was first obtained using the *bump* test. The plate was hit by a hammer and an estimate of the natural frequencies was obtained from the peaks in the subsequent frequency spectra. The peak-hold feature of the signal analyzer was used and an average from multiple readings was taken for each natural frequency. Later a sweep test around those averaged values was done to determine the exact natural frequencies. The first seven natural frequencies ω_i of the plate were found to be equal to 17.19 (first bending), 37.75 (first torsional), 110.06, 144.38, 155.56, 291.94, and 302.06 Hz. It is clearly evident that a few natural frequencies are nearly commensurate with other frequencies or a combination of some of them. So, we can expect some internal resonances to occur during a sweep of the excitation frequency around those natural frequencies.

6.2 Results

We performed three different frequency sweeps while keeping the value of the base excitation amplitude a_b fixed during each of those sweeps. In the first run, the excitation amplitude was held constant at 2.7g, while the excitation frequency Ω was varied around $\omega_1 + \omega_7$. In the second run, the excitation frequency was varied around ω_7 with $a_b = 4.5g$, and in the final run, a frequency sweep was made around ω_3 keeping the excitation amplitude fixed at 2g. Here, g ($= 9.8\ m/s^2$) denotes the acceleration due to gravity. For all the runs, we present the response frequency spectrum, and in some cases the input spectrum, response time trace, a pseudo-phase plane, or a Poincaré section is also presented. The frequency spectra of the signals obtained from the accelerometer and strain gages are calculated in real time using a flat-top window for the first two runs and a Hanning window for the third run.

6.2.1 RUN I: External Combination and Two-to-One Internal Resonances

While doing a backward frequency sweep, starting at $\Omega = 317$ Hz, we observed that the plate response changed from a periodic response with frequency Ω to one involving three distinct frequencies. The input and response frequency spectra at $\Omega = 316.81$ Hz are shown in Fig. 6.2. Basically there is a peak at the excitation frequency Ω and also near the two natural frequencies ω_1 (= 17.19 Hz) and ω_7 (= 302.06 Hz). Therefore, this is an example of an external combination resonance of the additive type ($\Omega \approx \omega_1 + \omega_7$), which occurs in the presence of quadratic nonlinearities. But it is well known that the nonlinearities present in the governing equations of an isotropic cantilever plate are cubic. We suspect that the initial curvature in the plate could be the cause of the quadratic nonlinearities present in the system. From Fig. 6.2, it is clear that the peaks at the natural frequencies ω_1 and ω_7 are much larger than the one at the excitation frequency Ω, with the peak at ω_1 being the largest. The large-amplitude first-mode response is also visible in the corresponding response time trace shown in Fig. 6.3. Thus, the external combination resonance has resulted in a transfer of energy from a low-amplitude, high-frequency mode to a high-amplitude, low-frequency mode. We note that a large static component is present in the plate responses as seen in the response spectrum (Fig. 6.2) and response time trace (Fig. 6.3). The cause for this large static component is the improper balancing of the Wheatstone bridges used in the measurement of the strain values.

In addition, we also obtained a Poincaré section using the two strain-gage signals as independent "pseudo" states (Nayfeh and Balachandran, 1995; Moon, 1987). Poincaré section is a powerful geometric tool used to analyze the dynamics of a system. Also, since the dynamics in a state space reconstructed from scalar time signals not representing displacement or velocity, like the strain-gage signals, is equivalent to the original dynamics, we conveniently use the strain-gage signals to obtain the Poincaré section. The two strain-gage signals were sampled at the excitation frequency Ω, and the resulting Poincaré section is shown in Fig. 6.4. Because the Poincaré section is essentially a closed curve, the plate motion is two-period quasiperiodic. The scatter in the results could be due to noise. Reducing the excitation frequency further to $\Omega = 315$ Hz resulted in the fourth and fifth modes getting excited, possibly through two-to-one internal resonances, as shown in Fig. 6.5. In addition to peaks at Ω, ω_1, and ω_7, we now also see peaks at $\omega_4 \approx \frac{1}{2}\omega_7$ and $\omega_5 \approx \frac{1}{2}\omega_7$. The two-to-one internal resonances are due to the presence of quadratic nonlinearities in the plate. The corresponding response time trace is shown in Fig. 6.6. Reducing the excitation frequency Ω gradually to 313.75 Hz resulted in a motion similar to the one shown in Fig. 6.2, and at $\Omega = 313.39$ Hz, the plate reverted back to the periodic

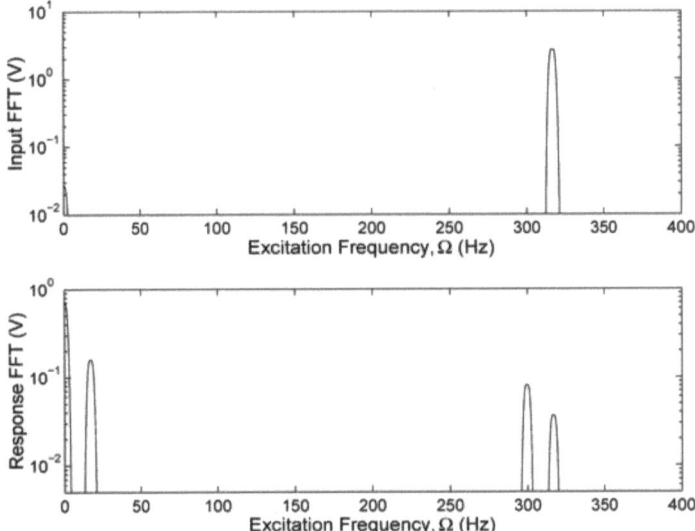

Figure 6.2: Input and response FFTs for $a_b = 2.7g$ and $\Omega = 316.81$ Hz.

motion with frequency Ω. During a forward sweep, we observed a behavior similar to the one seen during the backward frequency sweep.

6.2.2 RUN II: Two-to-One and Zero-to-One Internal Resonances

In the second run, the excitation amplitude a_b was held constant at $4.5g$ while the excitation frequency Ω was varied around the seventh natural frequency ω_7. During a backward sweep, starting at $\Omega = 312$ Hz, we observed a two-to-one internal resonance at $\Omega = 311.22$ Hz. The response spectrum at $\Omega = 311$ Hz is shown in Fig. 6.7. We see a large peak at the excitation frequency Ω and another one at $\frac{1}{2}\Omega$, which is close to the fifth natural frequency ω_5. In addition, we also obtained a pseudo-phase plane using the strain gage-one and accelerometer signals as independent states. The plot displayed in Fig. 6.8 shows a typical phase-plane trajectory for a system in which the response period is twice that of the excitation, as in the present case.

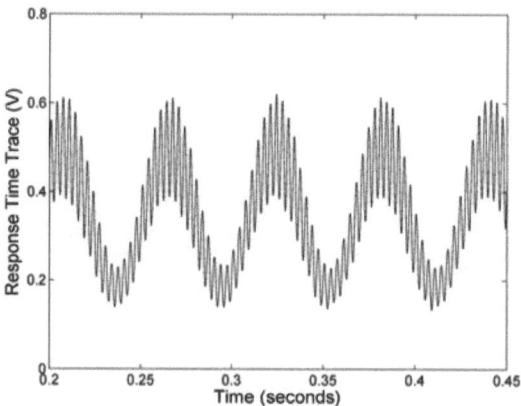

Figure 6.3: Response time trace for $a_b = 2.7g$ and $\Omega = 316.81$ Hz.

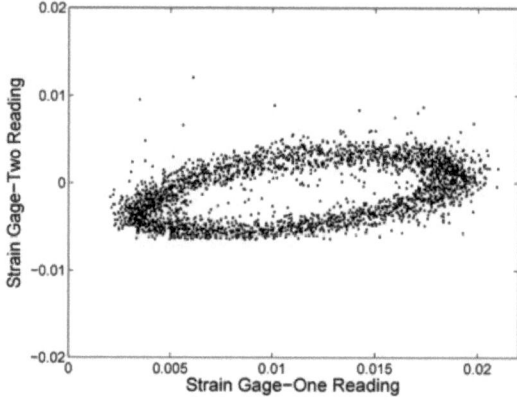

Figure 6.4: Poincaré section showing two-period quasiperiodic motion.

As Ω was reduced, the amplitude of the peak near ω_5 kept increasing, and at $\Omega = 305$ Hz it became larger than that of the peak at Ω. At $\Omega = 304.69$ Hz, the response of the fifth and seventh modes underwent a Hopf bifurcation, with the newly created frequency, also known as the modulation frequency, being close and slightly higher than the natural frequency of the first mode. Figure 6.9

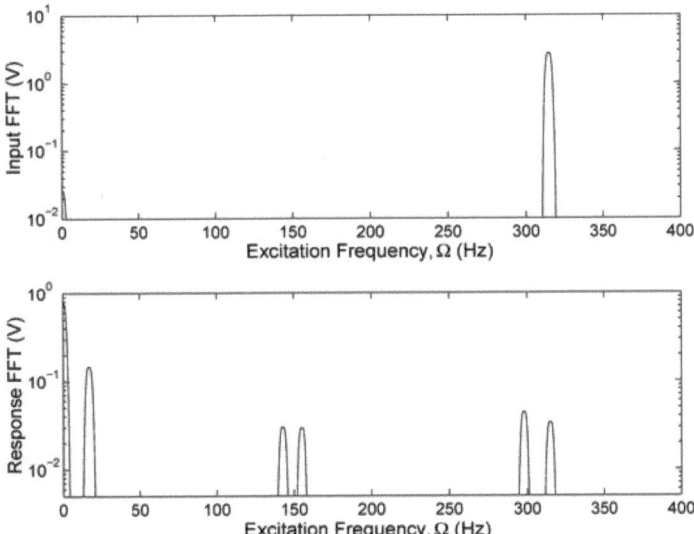

Figure 6.5: Input and response FFTs for $a_b = 2.7g$ and $\Omega = 315$ Hz.

shows the response spectrum at $\Omega = 304.5$ Hz, where the modulation frequency is equal to 17 Hz, which is close to ω_1 (= 17.19 Hz). It is also clear from the figure that the peak near the first mode frequency ω_1 is the largest. Thus, energy was transferred from two high-frequency modes, namely the fifth and seventh, to the low-frequency first mode via modulation. This is an example of the energy transfer between widely spaced modes, also known as zero-to-one internal resonance. As Ω was gradually reduced from 304.69 Hz, the modulation frequency tended to go towards the first natural frequency and the amplitude of the first-mode response kept increasing. But once the modulation frequency had passed the first natural frequency, the first-mode response amplitude started decreasing. At $\Omega = 304.41$ Hz, the plate response reverted back to the periodic motion with frequency Ω. A similar behavior was observed during a forward sweep of the excitation frequency in the frequency range of interest.

To study the influence of the excitation amplitude on the modulation frequency and, in turn, on the first-mode response, we reduced the excitation amplitude to $a_b = 3g$. At $\Omega = 300.94$ Hz, the

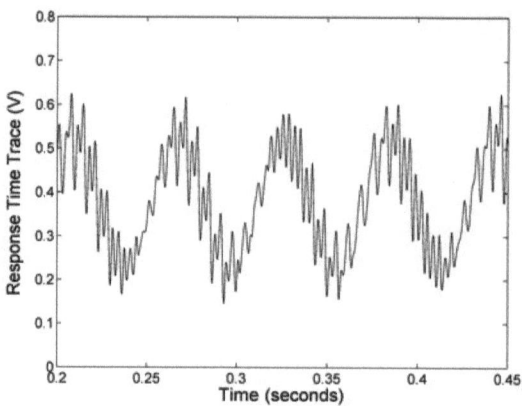

Figure 6.6: Response time trace for $a_b = 2.7g$ and $\Omega = 315$ Hz.

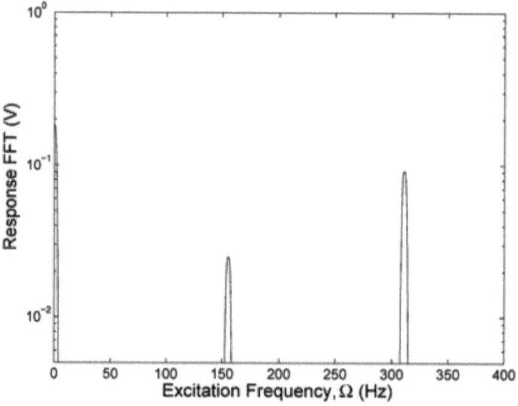

Figure 6.7: Response FFT for $a_b = 4.5g$ and $\Omega = 311$ Hz.

modulation frequency was equal to 11.7 Hz, which is far away from the first natural frequency ω_1 (= 17.19 Hz). Hence the amplitude of the first-mode response was very small, as is evident from the response spectrum shown in Fig. 6.10. A similar trend was observed in Chapter 5 while experimentally studying the transfer of energy between widely spaced modes in cantilever beams.

Chapter 6. Cantilever Plate 107

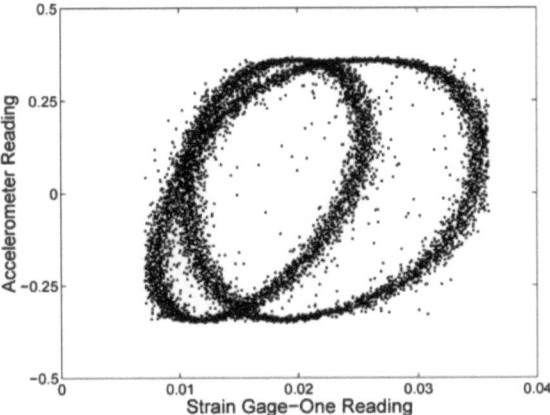

Figure 6.8: Pseudo-phase plane trajectory showing two-to-one internal resonance.

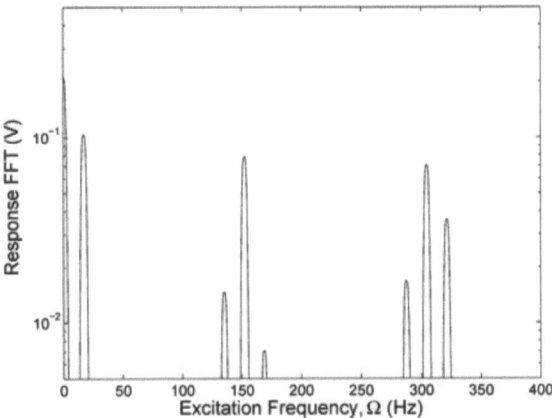

Figure 6.9: Response FFT for $a_b = 4.5g$ and $\Omega = 304.5$ Hz.

6.2.3 RUN III: Quasiperiodic Motion and Three-to-One Internal Resonance

This time we did a frequency sweep around the third-mode frequency ω_3 (= 110.06 Hz) keeping the base excitation fixed at $2g$. Starting with $\Omega = 110$ Hz and going backward in small increments, we

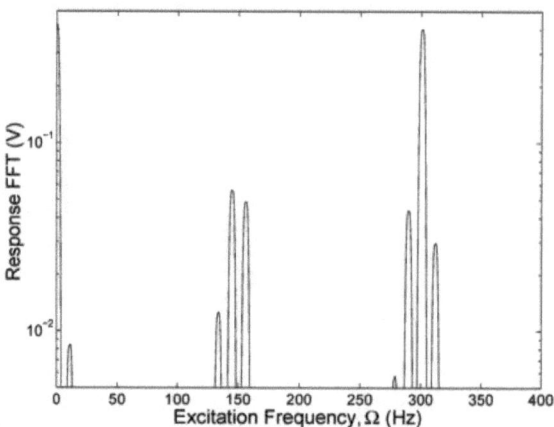

Figure 6.10: Response FFT for $a_b = 3g$ and $\Omega = 300.94$ Hz.

observed that the motion changed from a periodic motion to a two-period quasiperiodic motion at $\Omega = 109.42$ Hz. Basically there was a Hopf bifurcation leading to amplitude and phase modulation of the third-mode response. This is evident from the asymmetric sidebands observed around the peak at the excitation frequency in the response frequency spectrum as shown in Fig. 6.11, which is for the excitation frequency $\Omega = 109.41$ Hz. We observed no other peaks between 0 and Ω. As the excitation frequency was gradually reduced, we observed period-doubling bifurcations – first at $\Omega = 109.39$ Hz and then again at $\Omega = 109.35$ Hz. The FFTs of the corresponding zoomed-in responses are shown in Figs. 6.12 and 6.13. In Fig. 6.14, the complete input and response frequency spectra for $\Omega = 109.35$ Hz are shown. We see that now there is a small peak at the modulation frequency ($= 1.21$ Hz), which probably is the first-mode response. As the modulation frequency is very far away from the first natural frequency, the first-mode response amplitude is very small. Therefore, this might also possibly be an example of energy transfer between widely spaced modes. Furthermore, a close observation of the input spectrum reveals that the input is also modulated. This indicates a feedback from the structure to the shaker. A shaker-structure interaction is common and not unusual (McConnell, 1995).

Continuing with the frequency reduction, we observed that the plate motion goes back to the periodic motion with frequency Ω. But on further reduction, we observed a three-to-one internal resonance (starting at $\Omega = 108.61$ Hz), followed by a chaotically modulated motion (starting at $\Omega = $

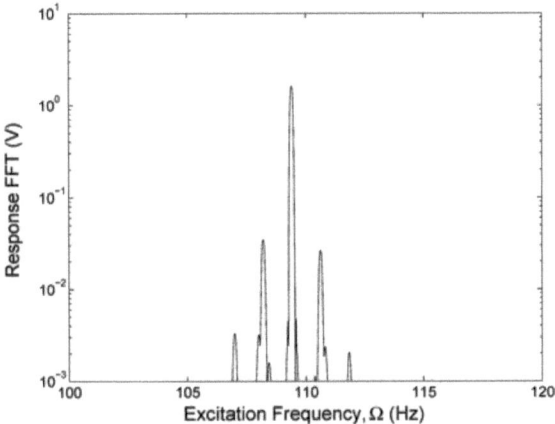

Figure 6.11: Response FFT for $a_b = 2g$ and $\Omega = 109.41$ Hz.

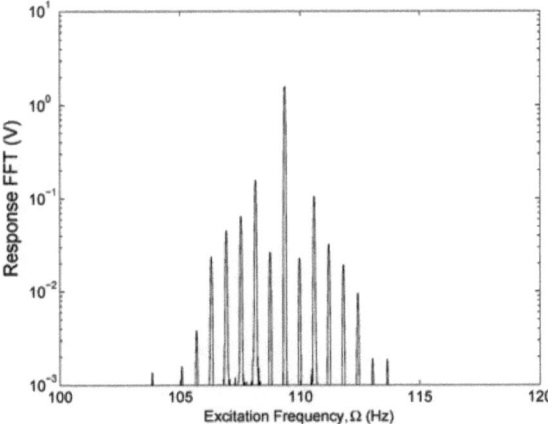

Figure 6.12: Response FFT for $a_b = 2g$ and $\Omega = 109.39$ Hz.

108.5 Hz), followed by a three-to-one internal resonance again (starting at $\Omega = 108.45$ Hz), and then the usual periodic motion (starting at $\Omega = 108.39$ Hz). The occurrence of the three-to-one internal

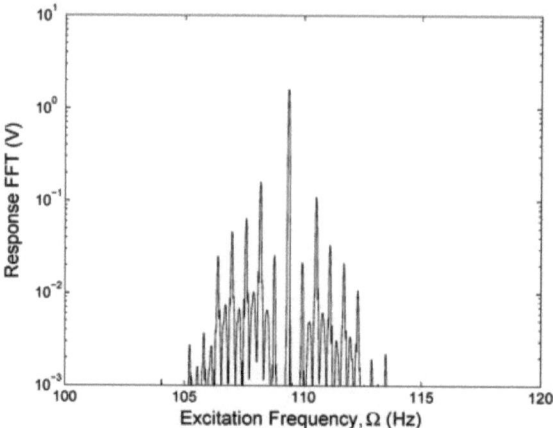

Figure 6.13: Response FFT for $a_b = 2g$ and $\Omega = 109.35$ Hz.

resonance is due to the fact that the plate system has inherent cubic geometric nonlinearities and because $\omega_3 \approx 3\omega_2$. The response spectra corresponding to the three-to-one internal resonance at $\Omega = 108.4$ Hz and the chaotically-modulated three-to-one internal resonance at $\Omega = 108.5$ Hz are shown in Figs. 6.15 and 6.16, respectively. The forward sweep also led to identical motions in the above frequency range of interest.

6.3 Closure

It is very interesting to see so many nonlinear dynamic phenomena in a simple aluminum plate. The initial curvature of the plate brought quadratic nonlinearities into the system, which provided an opportunity to observe phenomena which otherwise would not have occurred in a straight plate with just cubic nonlinearities. Also, many of the linear natural frequencies of the plate are nearly commensurate with one another, leading to multiple internal resonances.

The energy transfer from a low-amplitude, high-frequency mode to a high-amplitude, low-frequency mode via modulation was demonstrated. Also, the amount of energy transfer is found to be strongly dependent on the closeness of the modulation frequency to the natural frequency of the low-frequency

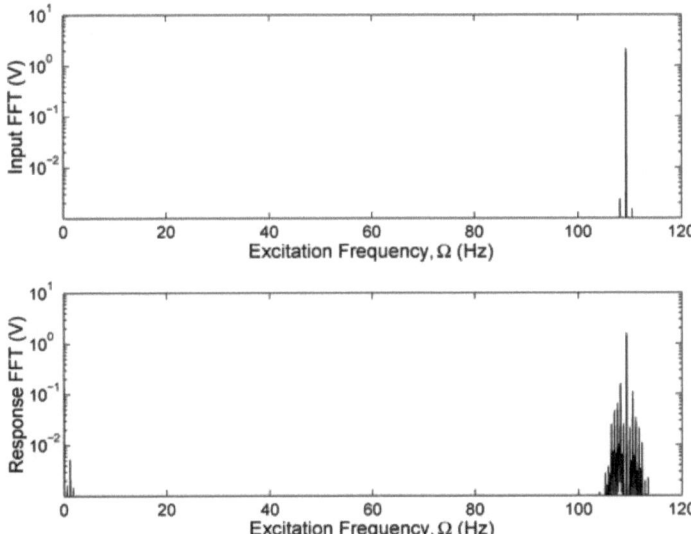

Figure 6.14: Input and response FFTs for $a_b = 2g$ and $\Omega = 109.35$ Hz.

mode. The results of RUN III show that the two-period quasiperiodic motion is due to an instability involving both the first and third modes. Probably it would not be amiss to generalize that a Hopf bifurcation in a structure under harmonic excitation would involve at least two modes. Hence, while trying to explain such a phenomenon analytically, a single-mode approximation should not be used as it might lead to erroneous results.

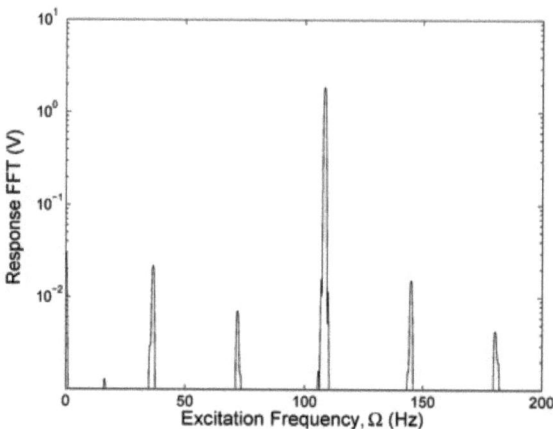

Figure 6.15: Response FFT for $a_b = 2g$ and $\Omega = 108.4$ Hz.

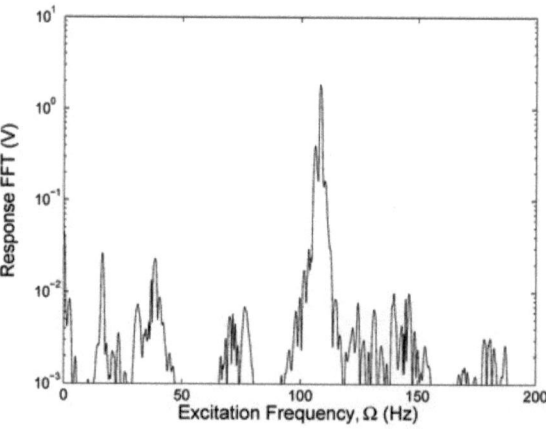

Figure 6.16: Response FFT for $a_b = 2g$ and $\Omega = 108.5$ Hz.

Chapter 7

Conclusions and Recommendation for Future Work

7.1 Summary

In this dissertation, we investigated the nonlinear vibrations of metallic cantilever beams and plates subjected to transverse harmonic excitations. In particular, we studied both experimentally and theoretically the energy transfer between widely spaced modes via modulation. Two cases – planar and nonplanar motions – were considered. In the planar case, a rectangular cross-section beam whose motion was restricted to the plane of excitation was studied. Whereas in the nonplanar case, a circular cross-section rod, where because of axial symmetry, one-to-one internal resonances occur at each natural frequency of the beam, was studied.

Experimentally it was observed that the transfer of energy between widely spaced modes is a function of the closeness of the modulation frequency to the natural frequency of the first mode. The modulation frequency, which depends on various parameters like the amplitude and frequency of excitation, damping factors, etc., has to be near the natural frequency of the first mode for significant transfer of energy from the directly excited high-frequency third mode to the low-frequency first mode. In such a case, visually we see a large swaying of the beam due to the large amplitude of the first-mode component in the beam response. This is very similar to primary resonance in a structural system where the closeness of an external excitation frequency to the natural frequencies of the structure dictates the magnitude of the structure response.

Based on the present observations, we can conclude that conventional methods used for reduction of response amplitude, such as decreasing amplitude of excitation or increasing damping, might in fact lead to an increase in the amplitude of the low-frequency response if they result in bringing the modulation frequency closer to the first natural frequency. Also, in most of the experiments dealing with energy transfer via modulation from high- to low-frequency modes, it was observed that the low-frequency mode is always the first mode. As the modulation frequency (or the newly created frequency) increases with increasing amplitude of excitation, it might be therefore possible to excite the second mode, instead of the first, if the forcing amplitude is large enough. However, this needs to be verified experimentally.

The reduced-order models were able to predict results qualitatively similar to the experimentally observed motions, but the results differed a lot quantitatively. Issues, like the shaker-structure interaction, nonlinear damping, initial curvature of the beam, extension of the beam axis, and inclusion of more modes in the modal discretization, were ignored while developing the models. It is, therefore, possible that one or more of these factors might be playing a bigger role than expected, and this needs to be further investigated.

An experimental parametric identification technique to estimate the linear and nonlinear damping coefficients and effective nonlinearity of a metallic cantilever beam was also developed. This method can be applied to any single-degree-of-freedom nonlinear system with weak cubic geometric and inertia nonlinearities. It was found that noninclusion of quadratic nonlinearity in the beam model might lead to incorrect parameter estimation. In addition, we proposed two methods, based on the elimination theory of polynomials, which can be used to determine both the critical forcing amplitude as well as the jump frequencies in single-degree-of-freedom nonlinear systems. These two methods have the potential of being applicable to multiple-degree-of-freedom nonlinear systems as well.

Finally, we carried out an experimental study of the response of a rectangular, aluminum cantilever plate to transverse harmonic excitations. We observed various nonlinear dynamic phenomena, like two-to-one and three-to-one internal resonances, external combination resonance, energy transfer between widely spaced modes via modulation, period-doubled motions, and chaos. The fact that a simple plate under harmonic excitation could display so many different nonlinear dynamic phenomena proves that nonlinear phenomena are ubiquitous. Some may however remark that these types of nonlinear resonances have very high activation thresholds; that is, the excitation level needs to be large to observe them. But since these unusual phenomena have been observed in real engineering structures and mechanical systems, there is a need to study such phenomena in more detail. For instance, the zero-to-one internal resonance was observed in the motion of the solar panels of the Hubble space telescope and also in the motion of the skin of a submarine. Many physical systems display nonlinear behavior, but only a watchful eye can recognize that. Most of the time our "linear" perspective (or in other words, our ignorance about nonlinear phenomena) makes us overlook any such *different* behavior in systems. It would be to our advantage if we start to think "nonlinearly".

7.2 Suggestions for Future Work

There is a lot of scope for improvement in the model used for describing the nonlinear, nonplanar dynamics of an isotropic, cantilever beam. Better models would lead to better agreement, both qualitatively and quantitatively, between the predicted and experimental results. Also, such models can be used to explain the many interesting motions observed experimentally.

Transfer of energy from high- to low-frequency modes is extremely dangerous as the amplitudes of the low-frequency modes are usually much larger than the directly-excited high-frequency modes. The strong dependence of the energy transfer between widely spaced modes on the modulation frequency can be utilized to control the amplitude of the low-frequency responses. Therefore, a study on such a control technique is highly recommended.

It would be interesting to see if energy transfer between widely spaced modes via modulation also occurs in micro-electromechanical systems (MEMS). In MEMS-based devices, the forcing term is highly nonlinear and varies with the displacement of the microstructure. One could study the influence of such a forcing term on the occurence of the energy transfer between widely spaced modes. Also, the applicability of the experimental parametric identification technique to estimate the damping factors and effective nonlinearity in such systems can be studied.

Bibliography

[1] ADAMS, R. D. AND BACON, D. G. C., *Measurement of the flexural damping capacity and dynamic Young's modulus of metals and reinforced plastics*, Journal of Physics D: Applied Physics, 6 (1973), pp. 27–41.

[2] ALHAZZA, K. A. AND NAYFEH, A. H., *Nonlinear vibrations of doubly-curved cross-ply shallow shells*, in Proceedings of the 42nd AIAA/ASME/ASCE/AHS/ASC Structures, Structural Dynamics, and Materials Conference, Seattle, Washington, 2001.

[3] ANAND, G. V., *Nonlinear resonance in stretched strings with viscous damping*, Journal of the Acoustical Society of America, 40 (1966), pp. 1517–1528.

[4] ANAND, G. V., *Stability of nonlinear oscillations of stretched strings*, Journal of the Acoustical Society of America, 46 (1969), pp. 667–677.

[5] ANDERSON, T. J., BALACHANDRAN, B., AND NAYFEH, A. H., *Observations of nonlinear interactions in a flexible cantilever beam*, in Proceedings of the 33rd AIAA Structures, Structural Dynamics, and Materials Conference, Dallas, Texas, 1992, pp. 1678–1686.

[6] ANDERSON, T. J., BALACHANDRAN, B., AND NAYFEH, A. H., *Nonlinear resonances in a flexible cantilever beam*, Journal of Vibration and Acoustics, 116 (1994), pp. 480–484.

[7] ANDERSON, T. J., NAYFEH, A. H., AND BALACHANDRAN, B., *Coupling between high-frequency modes and a low-frequency mode: Theory and experiment*, Nonlinear Dynamics, 11 (1996a), pp. 17–36.

[8] ANDERSON, T. J., NAYFEH, A. H., AND BALACHANDRAN, B., *Experimental verification of the importance of the nonlinear curvature in the response of a cantilever beam*, Journal of Vibration and Acoustics, 118 (1996b), pp. 21–27.

[9] ANNIGERI, B. S., CASSENTI, B. N., AND DENNIS, A. J., *Kinematics of small and large deformations of continua*, Engineering Computations, 2 (1985), pp. 247–256.

[10] ARAFAT, H. N., *Nonlinear Response of Cantilever Beams*, Ph.D. thesis, Virginia Polytechnic Institute and State University, Blacksburg, Virginia, 1999.

[11] ARAFAT, H. N. AND NAYFEH, A. H., *Transfer of energy from high- to low-frequency modes in the bending-torsion oscillation of cantilever beams*, in Proceedings of the 1999 ASME Design Engineering Technical Conferences, Las Vegas, Nevada, 1999.

[12] ARAFAT, H. N. AND NAYFEH, A. H., *The influence of nonlinear boundary conditions on the nonplanar autoparametric responses of an inextensible beam*, in Proceedings of the 2001 ASME Design Engineering Technical Conferences, Pittsburgh, Pennsylvania, 2001.

[13] ARAVAMUDAN, K. S. AND MURTHY, P. N., *Non-linear vibration of beams with time-dependent boundary conditions*, International Journal of Non-Linear Mechanics, 8 (1973), pp. 195–212.

[14] ATLURI, S., *Nonlinear vibrations of a hinged beam including nonlinear inertia effects*, Journal of Applied Mechanics, 40 (1973), pp. 121–126.

[15] BAILEY, C. D., *A new look at Hamilton's principle*, Foundations of Physics, 5 (1975a), pp. 433–451.

[16] BAILEY, C. D., *Application of Hamilton's law of varying action*, AIAA Journal, 13 (1975b), pp. 1154–1157.

[17] BENHAFSI, Y., PENNY, J. E. T., AND FRISWELL, M. I., *Identification of damping parameters of vibrating systems with cubic stiffness nonlinearity*, in Proceedings of the 13th International Modal Analysis Conference, Nashville, Tennessee, 1995, pp. 623–629.

[18] BOLOTIN, V. V., *The Dynamic Stability of Elastic Systems*, Holden-Day, San Francisco, California, 1964.

[19] BURGREEN, D., *Free vibrations of a pin-ended column with constant distance between pin ends*, Journal of Applied Mechanics, 18 (1951), pp. 135–139.

[20] BURTON, T. D. AND KOLOWITH, M., *Nonlinear resonances and chaotic motion in a flexible parametrically excited beam*, in Proceedings of the 2nd Conference on Nonlinear Vibrations, Stability, and Dynamics of Structures and Mechanisms, Blacksburg, Virginia, 1988.

[21] BUSBY, H. R. AND WEINGARTEN, V. I., *Non-linear response of a beam to periodic loading*, International Journal of Non-Linear Mechanics, 7 (1972), pp. 289–303.

[22] CAUGHEY, T. K. AND O'KELLY, M. E. J., *Effect of damping on the natural frequencies of linear dynamic systems*, Journal of the Acoustical Society of America, 33 (1961), pp. 1458–1461.

[23] CHIA, C. Y., *Nonlinear Analysis of Plates*, McGraw-Hill, New York, 1980.

[24] CHUN, K. R., *Free vibration of a beam with one end spring-hinged and the other free*, Journal of Applied Mechanics, 39 (1972), pp. 1154–1155.

[25] COLE, S. R., *Effects of spoiler surfaces on the aeroelastic behavior of a low-aspect-ratio rectangular wing*, AIAA Paper 90-0981, 1990.

[26] COWPER, G. R., *The shear coefficient in Timoshenko's beam theory*, Journal of Applied Mechanics, 33 (1966), pp. 335–340.

[27] COWPER, G. R., *On the accuracy of Timoshenko's beam theory*, Journal of the Engineering Mechanics Division, 94 (1968), pp. 1447–1453.

[28] COX, D., LITTLE, J., AND O'SHEA, D., *Ideals, Varieties, and Algorithms*, Springer-Verlag, New York, 1997.

[29] CRAWLEY, E. F. AND AUBERT, A. C., *Identification of nonlinear structural elements by force-state mapping*, AIAA Journal, 24 (1986), pp. 155–162.

[30] CRESPO DA SILVA, M. R. M., *Flexural-flexural oscillations of Beck's column subject to a planar harmonic excitation*, Journal of Sound and Vibration, 60 (1978a), pp. 133–144.

[31] CRESPO DA SILVA, M. R. M., *Harmonic non-linear response of Beck's column to a lateral excitation*, International Journal of Solids and Structures, 14 (1978b), pp. 987–997.

[32] CRESPO DA SILVA, M. R. M., *On the whirling of a base-excited cantilever beam*, Journal of the Acoustical Society of America, 67 (1980), pp. 704–707.

[33] CRESPO DA SILVA, M. R. M., *Non-linear flexural-flexural-torsional-extensional dynamics of beams. I: Formulation*, International Journal of Solids and Structures, 24 (1988a), pp. 1225–1234.

[34] CRESPO DA SILVA, M. R. M., *Non-linear flexural-flexural-torsional-extensional dynamics of beams. II: Response analysis*, International Journal of Solids and Structures, 24 (1988b), pp. 1235–1242.

[35] CRESPO DA SILVA, M. R. M., *Equations for nonlinear analysis of 3d motions of beams*, Applied Mechanics Reviews, 44 (1991), pp. S51–S59.

[36] CRESPO DA SILVA, M. R. M. AND GLYNN, C. C., *Nonlinear flexural-flexural-torsional dynamics of inextensional beams. I: Equations of motion*, Journal of Structural Mechanics, 6 (1978a), pp. 437–448.

[37] CRESPO DA SILVA, M. R. M. AND GLYNN, C. C., *Nonlinear flexural-flexural-torsional dynamics of inextensional beams. II: Forced motions*, Journal of Structural Mechanics, 6 (1978b), pp. 449–461.

[38] CRESPO DA SILVA, M. R. M. AND GLYNN, C. C., *Out-of-plane vibrations of a beam including non-linear inertia and non-linear curvature effects*, International Journal of Non-Linear Mechanics, 13 (1979a), pp. 261–271.

[39] CRESPO DA SILVA, M. R. M. AND GLYNN, C. C., *Non-linear non-planar resonant oscillations in fixed-free beams with support asymmetry*, International Journal of Solids and Structures, 15 (1979b), pp. 209–219.

[40] CRESPO DA SILVA, M. R. M. AND HODGES, D. H., *Nonlinear flexure and torsion of rotating beams, with application to helicopter rotor blades. I: Formulation*, Vertica, 10 (1986a), pp. 151–169.

[41] CRESPO DA SILVA, M. R. M. AND HODGES, D. H., *Nonlinear flexure and torsion of rotating beams, with application to helicopter rotor blades. II: Response and stability results*, Vertica, 10 (1986b), pp. 171–186.

[42] CRESPO DA SILVA, M. R. M. AND ZARETZKY, C. L., *Non-linear modal coupling in planar and non-planar responses of inextensional beams*, International Journal of Non-Linear Mechanics, 25 (1990), pp. 227–239.

[43] CRESPO DA SILVA, M. R. M. AND ZARETZKY, C. L., *Nonlinear flexural-flexural-torsional interactions in beams including the effect of torsional dynamics. I: Primary resonance*, Nonlinear Dynamics, 5 (1994), pp. 3–23.

[44] CRESPO DA SILVA, M. R. M., ZARETZKY, C. L., AND HODGES, D. H., *Effects of approximations on the static and dynamic response of a cantilever with a tip mass*, International Journal of Solids and Structures, 27 (1991), pp. 565–583.

[45] CUSUMANO, J. P. AND MOON, F. C., *Low dimensional behavior in chaotic nonplanar motions of a forced elastic rod: Experiment and theory*, in Nonlinear Dynamics in Engineering Systems, Schiehlen, W., ed., Springer-Verlag, New York, 1990, pp. 60–66.

[46] CUSUMANO, J. P. AND MOON, F. C., *Chaotic non-planar vibrations of the thin elastica. Part I: Experimental observation of planar instability*, Journal of Sound and Vibration, 179 (1995a), pp. 185–208.

[47] CUSUMANO, J. P. AND MOON, F. C., *Chaotic non-planar vibrations of the thin elastica. Part II: Derivation and analysis of a low-dimensional model*, Journal of Sound and Vibration, 179 (1995b), pp. 209–226.

[48] DANIELSON, D. A. AND HODGES, D. H., *Nonlinear beam kinematics by decomposition of the rotation tensor*, Journal of Applied Mechanics, 54 (1987), pp. 258–262.

[49] DOKUMACI, E., *Pseudo-coupled bending-torsion vibrations of beams under lateral parametric excitation*, Journal of Sound and Vibration, 58 (1978), pp. 233–238.

[50] DOUGHTY, T. A., DAVIES, P., AND BAJAJ, A. K., *A comparison of three techniques using steady state data to identify non-linear modal behavior of an externally excited cantilever beam*, Journal of Sound and Vibration, 249 (2002), pp. 785–813.

[51] DOWELL, E. H., *On asymptotic approximations to beam mode shapes*, Journal of Applied Mechanics, 51 (1984), p. 439.

[52] DOWELL, E. H., TRAYBAR, J., AND HODGES, D. H., *An experimental-theoretical correlation study of non-linear bending and torsion deformations of a cantilever beam*, Journal of Sound and Vibration, 50 (1977), pp. 533–544.

[53] DUGUNDJI, J., *Simple expressions for higher vibration modes of uniform Euler beams*, AIAA Journal, 26 (1988), pp. 1013–1014.

[54] DUGUNDJI, J. AND MUKHOPADHYAY, V., *Lateral bending-torsion vibrations of a thin beam under parametric excitation*, Journal of Applied Mechanics, 40 (1973), pp. 693–698.

[55] EISLEY, J. G., *Nonlinear vibration of beams and rectangular plates*, ZAMP, 15 (1964), pp. 167–175.

[56] EMAM, S. A. AND NAYFEH, A. H., *Nonlinear dynamics of a clamped-clamped buckled beam*, in Proceedings of the 43rd AIAA/ASME/ASCE/AHS/ASC Structures, Structural Dynamics, and Materials Conference, Denver, Colorado, 2002.

[57] EPSTEIN, M. AND MURRAY, D. W., *Three-dimensional large deformation analysis of thin walled beams*, International Journal of Solids and Structures, 12 (1976), pp. 867–876.

[58] EVAN-IWANOWSKI, R. M., *On the parametric response of structures*, Applied Mechanics Reviews, 18 (1965), pp. 699–702.

[59] EVAN-IWANOWSKI, R. M., *Resonance Oscillations in Mechanical Systems*, Elsevier, New York, 1976.

[60] EVENSEN, D. A., *Nonlinear vibrations of beams with various boundary conditions*, AIAA Journal, 6 (1968), pp. 370–372.

[61] FAHEY, S. O. AND NAYFEH, A. H., *Experimental nonlinear identification of a single structural mode*, in Proceedings of the 16th International Modal Analysis Conference, Orlando, Florida, 1998, pp. 737–745.

[62] FENG, Z. C., *Nonresonant modal interactions*, in Proceedings of the 1995 ASME Design Engineering Technical Conferences, Boston, Massachusetts, 1995, pp. 511–517.

[63] FRISWELL, M. I. AND PENNY, J. E. T., *The accuracy of jump frequencies in series solutions of the response of a Duffing oscillator*, Journal of Sound and Vibration, 169 (1994), pp. 261–269.

[64] FUNG, Y. C., *Foundations of Solid Mechanics*, Prentice-Hall, Englewood Cliffs, New Jersey, 1965.

[65] GINSBERG, J. H., *The effects of damping on a non-linear system with two degrees of freedom*, International Journal of Non-Linear Mechanics, 7 (1972), pp. 323–336.

[66] GORMAN, D. J., *Free Vibration Analysis of Beams and Shafts*, Wiley, New York, 1975.

[67] GRIFFITHS, L. W., *Introduction to the Theory of Equations*, Wiley, New York, 1947.

[68] HADDOW, A. G. AND HASAN, S. M., *Nonlinear oscillation of a flexible cantilever: Experimental results*, in Proceedings of the 2nd Conference on Nonlinear Vibrations, Stability, and Dynamics of Structures and Mechanisms, Blacksburg, Virginia, 1988.

[69] HAIGHT, E. C. AND KING, W. W., *Stability of parametrically excited vibrations of an elastic rod*, Developments in Theoretical and Applied Mechanics, 5 (1971), pp. 677–713.

[70] HAIGHT, E. C. AND KING, W. W., *Stability of nonlinear oscillations of an elastic rod*, Journal of the Acoustical Society of America, 52 (1972), pp. 899–911.

[71] HALLER, G., *Chaos Near Resonance*, Springer-Verlag, New York, 1999.

[72] HENRY, R. F. AND TOBIAS, S. A., *Modes at rest and their stability in coupled non-linear systems*, Journal of Mechanical Engineering Science, 3 (1961), pp. 163–173.

[73] HINNANT, H. E. AND HODGES, D. H., *Nonlinear analysis of a cantilever beam*, AIAA Journal, 26 (1988), pp. 1521–1527.

[74] HO, C.-H., SCOTT, R. A., AND EISLEY, J. G., *Non-planar, non-linear oscillations of a beam. I: Forced motions*, International Journal of Non-Linear Mechanics, 10 (1975), pp. 113–127.

[75] HO, C.-H., SCOTT, R. A., AND EISLEY, J. G., *Non-planar, non-linear oscillations of a beam. II: Free motions*, Journal of Sound and Vibration, 47 (1976), pp. 333–339.

[76] HODGES, D. H., *Finite rotation and nonlinear beam kinematics*, Vertica, 11 (1987a), pp. 297–307.

[77] HODGES, D. H., *Nonlinear beam kinematics for small strains and finite rotations*, Vertica, 11 (1987b), pp. 573–589.

[78] HODGES, D. H., CRESPO DA SILVA, M. R. M., AND PETERS, D. A., *Nonlinear effects in the static and dynamic behavior of beams and rotor blades*, Vertica, 12 (1988), pp. 243–256.

[79] HODGES, D. H. AND DOWELL, E. H., *Nonlinear equations of motion for the elastic bending and torsion of twisted nonuniform rotor blades*, NASA Technical Note D-7818, 1974.

[80] HUANG, T. C., *The effect of rotatory inertia and of shear deformation on the frequency and normal mode equations of uniform beams with simple end conditions*, Journal of Applied Mechanics, 28 (1961), pp. 579–584.

[81] HUGHES, G. C. AND BERT, C. W., *Effect of gravity on nonlinear oscillations of a horizontal, immovable-end beam*, in Proceedings of the 3rd Conference on Nonlinear Vibrations, Stability, and Dynamics of Structures and Mechanisms, Blacksburg, Virginia, 1990.

[82] HYER, M. W., *Whirling of a base-excited cantilever beam*, Journal of the Acoustical Society of America, 65 (1979), pp. 931–939.

[83] IBRAHIM, R. A. AND HIJAWI, M., *Deterministic and stochastic response of nonlinear coupled bending-torsion modes in a cantilever beam*, Nonlinear Dynamics, 16 (1998), pp. 259–292.

[84] KANE, T. R., RYAN, R. R., AND BANERJEE, A. K., *Dynamics of a cantilever beam attached to a moving base*, Journal of Guidance, 10 (1987), pp. 139–151.

[85] KAPANIA, R. K. AND PARK, S., *Parametric identification of nonlinear structural dynamic systems using time finite element method*, AIAA Journal, 35 (1997), pp. 719–726.

[86] KÁRMÁN, T. V., *The engineer grapples with nonlinear problems*, Bulletin of the American Mathematical Society, 46 (1940), pp. 615–683.

[87] KRAUSS, R. W. AND NAYFEH, A. H., *Comparison of experimental identification techniques for a nonlinear SDOF system*, in Proceedings of the 17th International Modal Analysis Conference, Santa Barbara, California, 1999a, pp. 1182–1187.

[88] KRAUSS, R. W. AND NAYFEH, A. H., *Experimental nonlinear identification of a single mode of a transversely excited beam*, Nonlinear Dynamics, 18 (1999b), pp. 69–87.

[89] LACARBONARA, W., *Direct treatment and discretizations of non-linear spatially continuous systems*, Journal of Sound and Vibration, 221 (1999), pp. 849–866.

[90] LANGFORD, W., *Normal form analysis of Nayfeh's abnormal resonance*, in Annual Meeting of the Canadian Applied and Industrial Mathematics Society, Victoria, Canada, 2001.

[91] LEE, G.-M., *Estimation of non-linear system parameters using higher-order frequency response functions*, Mechanical Systems and Signal Processing, 11 (1997), pp. 219–228.

[92] LOVE, A. E. H., *A Treatise on the Mathematical Theory of Elasticity*, Dover, New York, 1944.

[93] MALATKAR, P. AND NAYFEH, A. H., *Calculation of the jump frequencies in the response of s.d.o.f. non-linear systems*, Journal of Sound and Vibration, 254 (2002), pp. 1005–1011.

[94] MALATKAR, P. AND NAYFEH, A. H., *A parametric identification technique for SDOF weakly nonlinear systems with cubic nonlinearities*, Journal of Vibration and Control, 9 (2003a), pp. 317–336.

[95] MALATKAR, P. AND NAYFEH, A. H., *On the transfer of energy between widely spaced modes in structures*, Nonlinear Dynamics, 31 (2003b), pp. 225–242.

[96] MALATKAR, P. AND NAYFEH, A. H., *A plethora of nonlinear dynamics phenomena observed in a simple cantilever plate*, 2003 ASME Design Engineering Technical Conferences, Chicago, Illinois, accepted (2003c).

[97] MASRI, S. F. AND CAUGHEY, T. K., *A nonparametric identification technique for nonlinear dynamic problems*, Journal of Applied Mechanics, 46 (1979), pp. 433–447.

[98] MASRI, S. F., SASSI, H., AND CAUGHEY, T. K., *Nonparametric identification of nearly arbitrary nonlinear systems*, Journal of Applied Mechanics, 49 (1982), pp. 619–628.

[99] MCCONNELL, K. G., *Vibration Testing: Theory and Practice*, Wiley, New York, 1995.

[100] MCDANIEL, J. G., WIDJAJA, F., BARBONE, P. E., AND PIERCE, A. D., *Estimating natural frequencies and mode shapes from forced response calculations*, AIAA Journal, 40 (2002), pp. 758–764.

[101] MCDONALD, P. H., *Nonlinear dynamic coupling in a beam vibration*, Journal of Applied Mechanics, 22 (1955), pp. 573–578.

[102] MEIROVITCH, L., *Analytical Methods in Dynamics*, Macmillan, New York, 1967.

[103] MEIROVITCH, L., *Principles and Techniques of Vibrations*, Prentice Hall, Upper Saddle River, New Jersey, 1997.

[104] MILES, J. W., *Stability of forced oscillations of a vibrating string*, Journal of the Acoustical Society of America, 38 (1965), pp. 855–861.

[105] MINDLIN, R. D., *Influence of rotatory inertia and shear on flexural motions of isotropic, elastic plates*, Journal of Applied Mechanics, 18 (1951), pp. 31–38.

[106] MOHAMMAD, K. S., WORDEN, K., AND TOMLINSON, G. R., *Direct parameter estimation for linear and non-linear structures*, Journal of Sound and Vibration, 152 (1992), pp. 471–499.

[107] MOON, F. C., *Chaotic Vibrations – An Introduction for Applied Scientists and Engineers*, Wiley, New York, 1987.

[108] MURTHY, G. S. S. AND RAMAKRISHNA, B. S., *Nonlinear character of resonance in stretched strings*, Journal of the Acoustical Society of America, 38 (1965), pp. 461–471.

[109] MURTY, A. V. K., *Vibrations of short beams*, AIAA Journal, 8 (1970), pp. 34–38.

[110] NAYFEH, A. H., *Nonlinear transverse vibrations of beams with properties that vary along the length*, Journal of the Acoustical Society of America, 53 (1973a), pp. 766–770.

[111] NAYFEH, A. H., *Perturbation Methods*, Wiley, New York, 1973b.

[112] NAYFEH, A. H., *Introduction to Perturbation Techniques*, Wiley, New York, 1981.

[113] NAYFEH, A. H., *Parametric identification of nonlinear dynamic systems*, Computers & Structures, 20 (1985), pp. 487–493.

[114] NAYFEH, A. H., *Perturbation methods in nonlinear dynamics*, in Nonlinear Dynamics Aspects of Particle Accelerators, Jowett, J. M., Month, M., and Turner, S., eds., Springer-Verlag, Berlin, Germany, 1986, pp. 238–314.

[115] NAYFEH, A. H., *Reduced-order models of weakly nonlinear spatially continuous systems*, Nonlinear Dynamics, 16 (1998), pp. 105–125.

[116] NAYFEH, A. H., *Nonlinear Interactions*, Wiley, New York, 2000.

[117] NAYFEH, A. H. AND ARAFAT, H. N., *An overview of nonlinear system dynamics*, in Structural Dynamics in 2000: Current Status and Future Directions, Ewins, D. J. and Inman, D. J., eds., Research Studies Press, 2001, pp. 225–256.

[118] NAYFEH, A. H. AND ARAFAT, H. N., *Nonlinear responses of suspended cables to primary resonance excitations*, in Proceedings of the 43rd AIAA/ASME/ASCE/AHS/ASC Structures, Structural Dynamics, and Materials Conference, Denver, Colorado, 2002.

[119] NAYFEH, A. H. AND BALACHANDRAN, B., *Modal interactions in dynamical and structural systems*, Applied Mechanics Reviews, 42 (1989), pp. S175–S201.

[120] NAYFEH, A. H. AND CHIN, C.-M., *Nonlinear interactions in a parametrically excited system with widely spaced frequencies*, Nonlinear Dynamics, 7 (1995), pp. 195–216.

[121] NAYFEH, A. H. AND CHIN, C.-M., *Perturbation Methods with Mathematica*, Dynamic Press, Virginia, 1999; http://www.esm.vt.edu/~anayfeh.

[122] NAYFEH, A. H. AND LACARBONARA, W., *On the discretization of distributed-parameter systems with quadratic and cubic nonlinearities*, Nonlinear Dynamics, 13 (1997), pp. 203–220.

[123] NAYFEH, A. H. AND MOOK, D. T., *Nonlinear Oscillations*, Wiley, New York, 1979.

[124] NAYFEH, A. H. AND MOOK, D. T., *Energy transfer from high-frequency to low-frequency modes in structures*, Journal of Vibration and Acoustics, 117 (1995), pp. 186–195.

[125] NAYFEH, A. H., MOOK, D. T., AND LOBITZ, D. W., *Numerical-perturbation method for the nonlinear analysis of structural vibrations*, AIAA Journal, 12 (1974), pp. 1222–1228.

[126] NAYFEH, A. H., MOOK, D. T., AND SRIDHAR, S., *Nonlinear analysis of the forced response of structural elements*, Journal of the Acoustical Society of America, 55 (1974), pp. 281–291.

[127] NAYFEH, A. H. AND PAI, P. F., *Linear and Nonlinear Structural Mechanics*, Wiley, New York, 2003 (in preparation).

[128] NAYFEH, S. A., *Nonlinear Dynamics of Systems Involving Widely Spaced Frequencies*, Master's thesis, Virginia Polytechnic Institute and State University, Blacksburg, Virginia, 1993.

[129] NAYFEH, S. A. AND NAYFEH, A. H., *Nonlinear interactions between two widely spaced modes – external excitation*, International Journal of Bifurcation and Chaos, 3 (1993), pp. 417–427.

[130] NAYFEH, S. A. AND NAYFEH, A. H., *Energy transfer from high- to low-frequency modes in a flexible structure via modulation*, Journal of Vibration and Acoustics, 116 (1994), pp. 203–207.

[131] NORDGREN, R. P., *On computation of the motion of elastic rods*, Journal of Applied Mechanics, 41 (1974), pp. 777–780.

[132] OH, K., *A Theoretical and Experimental Study of Modal Interactions in Metallic and Laminated Composite Plates*, Ph.D. thesis, Virginia Polytechnic Institute and State University, Blacksburg, Virginia, 1994.

[133] OH, K. AND NAYFEH, A. H., *High- to low-frequency modal interactions in a cantilever composite plate*, Journal of Vibration and Acoustics, 120 (1998), pp. 579–587.

[134] OSTIGUY, G. L. AND EVAN-IWANOWSKI, R. M., *On dynamic stability and nonlinear modal interaction of parametrically excited rectangular plates*, in Dynamics and Vibration of Time-Varying Systems and Structures, Sinha, S. C. and Evan-Iwanowski, R. M., eds., 1993, pp. 465–474.

[135] PAI, P. F., *Nonlinear Flexural-Flexural-Torsional Dynamics of Metallic and Composite Beams*, Ph.D. thesis, Virginia Polytechnic Institute and State University, Blacksburg, Virginia, 1990.

[136] PAI, P. F. AND NAYFEH, A. H., *Three-dimensional nonlinear vibrations of composite beams. I: Equations of motion*, Nonlinear Dynamics, 1 (1990a), pp. 477–502.

[137] PAI, P. F. AND NAYFEH, A. H., *Non-linear non-planar oscillations of a cantilever beam under lateral base excitations*, International Journal of Non-Linear Mechanics, 25 (1990b), pp. 455–474.

[138] PAI, P. F. AND NAYFEH, A. H., *A nonlinear composite beam theory*, Nonlinear Dynamics, 3 (1992), pp. 273–303.

[139] PAI, P. F. AND NAYFEH, A. H., *A fully nonlinear theory of curved and twisted composite rotor blades accounting for warpings and three-dimensional stress effects*, International Journal of Solids and Structures, 31 (1994), pp. 1309–1340.

[140] PAKDEMIRLI, M., NAYFEH, S. A., AND NAYFEH, A. H., *Analysis of one-to-one autoparametric resonances in cables – discretization vs. direct treatment*, Nonlinear Dynamics, 8 (1995), pp. 65–83.

[141] POPOVIC, P., NAYFEH, A. H., OH, K., AND NAYFEH, S. A., *An experimental investigation of energy transfer from a high-frequency mode to a low-frequency mode in a flexible structure*, Journal of Vibration and Control, 1 (1995), pp. 115–128.

[142] RAO, G. V., RAJU, I. S., AND RAJU, K. K., *Nonlinear vibrations of beams considering shear deformation and rotary inertia*, AIAA Journal, 14 (1976), pp. 685–687.

[143] RAY, J. D. AND BERT, C. W., *Nonlinear vibrations of a beam with pinned ends*, Journal of Engineering for Industry, 91 (1969), pp. 997–1004.

[144] REGA, G., LACARBONARA, W., NAYFEH, A. H., AND CHIN, C.-M., *Multiple resonances in suspended cables: Direct versus reduced-order models*, International Journal of Non-Linear Mechanics, 34 (1999), pp. 901–924.

[145] ROSEN, A. AND FRIEDMANN, P., *The nonlinear behavior of elastic slender straight beams undergoing small strains and moderate rotations*, Journal of Applied Mechanics, 46 (1979), pp. 161–168.

[146] ROSEN, A., LOEWY, R. G., AND MATHEW, M. B., *Nonlinear analysis of pretwisted rods using "principal curvature transformation." Part I: Theoretical derivation*, AIAA Journal, 25 (1987a), pp. 470–478.

[147] ROSEN, A., LOEWY, R. G., AND MATHEW, M. B., *Nonlinear analysis of pretwisted rods using "principal curvature transformation." Part II: Numerical results*, AIAA Journal, 25 (1987b), pp. 598–604.

[148] ROSEN, A., LOEWY, R. G., AND MATHEW, M. B., *Nonlinear dynamics of slender rods*, AIAA Journal, 25 (1987c), pp. 611–619.

[149] ROSENBERG, R. M., *Nonlinear oscillations*, Applied Mechanics Reviews, 14 (1961), pp. 837–841.

[150] SATHYAMOORTHY, M., *Nonlinear analysis of beams. Part I: A survey of recent advances*, Shock and Vibration Digest, 14 (1982a), pp. 19–35.

[151] SATHYAMOORTHY, M., *Nonlinear analysis of beams. Part II: Finite element methods*, Shock and Vibration Digest, 14 (1982b), pp. 7–18.

[152] SATHYAMOORTHY, M., *Nonlinear Analysis of Structures*, CRC Press, New York, 1997.

[153] SATO, K., SAITO, H., AND OTOMI, K., *The parametric response of a horizontal beam carrying a concentrated mass under gravity*, Journal of Applied Mechanics, 45 (1978), pp. 643–648.

[154] SHAMES, I. H. AND DYM, C. L., *Energy and Finite Element Methods in Structural Mechanics*, McGraw-Hill, New York, 1985.

[155] SHYU, I.-M. K., MOOK, D. T., AND PLAUT, R. H., *Whirling of a forced cantilevered beam with static deflection. I: Primary resonance*, Nonlinear Dynamics, 4 (1993a), pp. 227–249.

[156] SHYU, I.-M. K., MOOK, D. T., AND PLAUT, R. H., *Whirling of a forced cantilevered beam with static deflection. II: Superharmonic and subharmonic resonances*, Nonlinear Dynamics, 4 (1993b), pp. 337–356.

[157] SHYU, I.-M. K., MOOK, D. T., AND PLAUT, R. H., *Whirling of a forced cantilevered beam with static deflection. III: Passage through resonance*, Nonlinear Dynamics, 4 (1993c), pp. 461–481.

[158] SINCLAIR, G. B., *The non-linear bending of a cantilever beam with shear and longitudinal deformations*, International Journal of Non-Linear Mechanics, 14 (1979), pp. 111–122.

[159] SMITH, S. W., BALACHANDRAN, B., AND NAYFEH, A. H., *Nonlinear interactions and the Hubble space telescope*, in Proceedings of the 33rd AIAA Structures, Structural Dynamics, and Materials Conference, Paper 92-4617, Dallas, Texas, 1992, pp. 1460–1469.

[160] STRUTT, J. W., *The Theory of Sound*, Volume I, Dover, New York, 1945.

[161] TABADDOR, M., *Influence of nonlinear boundary conditions on the single-mode response of a cantilever beam*, International Journal of Solids and Structures, 37 (2000), pp. 4915–4931.

[162] TABADDOR, M. AND NAYFEH, A. H., *An experimental investigation of multimode responses in a cantilever beam*, Journal of Vibration and Acoustics, 119 (1997), pp. 532–538.

[163] TEZAK, E. G., MOOK, D. T., AND NAYFEH, A. H., *Nonlinear analysis of the lateral response of columns to periodic loads*, Journal of Mechanical Design, 100 (1978), pp. 651–659.

[164] TIMOSHENKO, S. P., *On the correction for shear of the differential equation for transverse vibrations of prismatic bars*, Philosophical Magazine and Journal of Science, 41 (1921), pp. 744–747.

[165] TIMOSHENKO, S. P., *On the transverse vibrations of bars of uniform cross-section*, Philosophical Magazine and Journal of Science, 43 (1922), pp. 125–131.

[166] TIMOSHENKO, S. P., *History of Strength of Materials*, Dover, New York, 1983.

[167] TIMOSHENKO, S. P. AND GOODIER, J. N., *Theory of Elasticity*, McGraw-Hill, New York, 1970.

[168] TIMOSHENKO, S. P. AND YOUNG, D. H., *Elements of Strength of Materials*, Litton Educational Publishing, New York, 1968.

[169] TONDL, A., *The application of skeleton curves and limit envelopes to analysis of nonlinear vibration*, Shock and Vibration Digest, 7 (1975), pp. 3–20.

[170] TSO, W. K., *Parametric torsional stability of a bar under axial excitation*, Journal of Applied Mechanics, 35 (1968), pp. 13–19.

[171] WAGNER, H. AND RAMAMURTI, V., *Beam vibrations – a review*, Shock and Vibration Digest, 9 (1977), pp. 17–24.

[172] WEE, C. E. AND GOLDMAN, R., *Elimination and resultants. part 1: Elimination and bivariate resultants*, IEEE Computer Graphics and Applications, 15 (1995a), pp. 69–77.

[173] WEE, C. E. AND GOLDMAN, R., *Elimination and resultants. part 2: Multivariate resultants*, IEEE Computer Graphics and Applications, 15 (1995b), pp. 60–69.

[174] WEMPNER, G., *Mechanics of Solids with Applications to Thin Bodies*, Sijthoff and Noordhoff, Rockville, Maryland, 1981.

[175] WOINOWSKY-KRIEGER, S., *The effect of an axial force on the vibration of hinged bars*, Journal of Applied Mechanics, 17 (1950), pp. 35–36.

[176] WOLFRAM, S., *The Mathematica Book*, Cambridge University Press, New York, 1999.

[177] WORDEN, K., *On jump frequencies in the response of the Duffing oscillator*, Journal of Sound and Vibration, 198 (1996), pp. 522–525.

[178] WORDEN, K. AND TOMLINSON, G. R., *Nonlinearity in Structural Dynamics*, Institute of Physics, Philadelphia, Pennsylvania, 2001.

[179] YAMAKI, N. AND CHIBA, M., *Nonlinear vibrations of a clamped rectangular plate with initial deflection and initial edge displacement – Part I: Theory*, Thin-Walled Structures, 1 (1983), pp. 3–29.

[180] YAMAKI, N., OTOMO, K., AND CHIBA, M., *Nonlinear vibrations of a clamped rectangular plate with initial deflection and initial edge displacement – Part II: Experiment*, Thin-Walled Structures, 1 (1983), pp. 101–119.

[181] YASUDA, K. AND KAMIYA, K., *Experimental identification technique of nonlinear beams in time domain*, Nonlinear Dynamics, 18 (1999), pp. 185–202.

[182] YASUDA, K., KAMIYA, K., AND KOMAKINE, M., *Experimental identification technique of vibrating structures with geometrical nonlinearity*, Journal of Applied Mechanics, 64 (1997), pp. 275–280.

[183] ZARETZKY, C. L. AND CRESPO DA SILVA, M. R. M., *Experimental investigation of non-linear modal coupling in the response of cantilever beams*, Journal of Sound and Vibration, 174 (1994a), pp. 145–167.

[184] ZARETZKY, C. L. AND CRESPO DA SILVA, M. R. M., *Nonlinear flexural-flexural-torsional interactions in beams including the effect of torsional dynamics. II: Combination resonance*, Nonlinear Dynamics, 5 (1994b), pp. 161–180.

[185] ZAVODNEY, L. D., *Can the modal analyst afford to be ignorant of nonlinear vibration phenomena?*, in Proceedings of the 5th International Modal Analysis Conference, London, 1987, pp. 154–159.